THE EVOLUTION OF
THEODOSIUS DOBZHANSKY

The Evolution of
Theodosius Dobzhansky

*Essays on His Life and Thought
in Russia and America*

MARK B. ADAMS

EDITOR

PRINCETON UNIVERSITY PRESS

PRINCETON, NEW JERSEY

Library of Congress Cataloging-in-Publication Data

The Evolution of Theodosius Dobzhansky : essays on
his life and thought in Russia and America /
Mark B. Adams, editor.
p. cm.
Selected papers from the International Symposium on
Theodosius Dobzhansky, held in Leningrad Sept. 17–19, 1990
Includes bibliographical references.
ISBN 0-691-03479-6
1. Dobzhansky, Theodosius Grigorievich, 1900–1975—
Congresses. 2. Dobzhansky, Theodosius Grigorievich,
1900–1975—Political and social views—Congresses.
3. Genetics—History—Congresses. 4. Genetics—
Soviet Union—History—Congresses. 5. Evolution
(Biology)—History—Congresses. I. Adams, Mark B.
II. International Symposium on Theodosius Dobzhansky
(1900 : Saint Petersburg, Russia)
QH31.D58E96 1994
575.1′092—dc20   93-42144
[B]

This book has been composed in Adobe Utopia

Princeton University Press books are printed on
acid-free paper and meet the guidelines for
permanence and durability of the Committee on
Production Guidelines for Book Longevity
of the Council on Library Resources

Printed in the United States of America

2   4   6   8   10   9   7   5   3   1

CONTENTS

PREFACE

ORGANIZING and editing this volume on the life and thought of Theodosius Dobzhansky has been a special privilege. In some respects Dobzhansky was indirectly responsible for my own lifelong interest in evolutionary theory, genetics, and the social implications of science: my library still contains the dog-eared copies of *Evolution, Genetics, and Man* and *Heredity, Race, and Society* that I pored over as a student in junior high school in the late 1950s.

I first met Dobzhansky in 1974 at Ernst Mayr's conference on the evolutionary synthesis at the American Academy of Arts and Sciences. At the time I was investigating Russian genetics during the 1920s. At the gracious invitation of Doby (as he insisted on being called), I taped a week's worth of interviews with him at Mather Camp in California in July 1974. The interviews were conducted in English and Russian and concerned the years before Dobzhansky came to the United States. Speaking in Russian, he gave a richer and somewhat different account of those years than appears in his English works and memoirs. When we returned to Davis, Dobzhansky gave me copies of his Russian correspondence (among the only correspondence he did not throw away during the move from Columbia to Rockefeller University), including many long, detailed letters from his mentor, Filipchenko, which he had received during his very first years in the States (1927–1930). When I asked Dobzhansky where his letters to Filipchenko were, he indicated that he thought they had probably been destroyed during the Leningrad blockade or because of Lysenkoism.

In a convoluted sort of way, that offhand remark was the seed that grew into this volume. During the academic year 1976–1977, some two years after Doby's death, I was in the Soviet Union doing research on the history of Russian population genetics. Russia was then a very different country. Although Lysenko had lost his hegemony over Soviet biology a decade earlier (he died in November 1976 during my stay), the Soviet Union was still a remarkably closed society. All the historical archives I wished to examine had to be pre-inspected by my official *konsul'tant*. The rationale was that a scholar from a "capitalist" country should not be able to see scientific "secrets," anything remotely political, or anything that might conceivably reflect badly on the Soviet Union. I could not see anything from the archives that had not been preapproved by my "consultant," and the upshot was that, with few exceptions, I was permitted to see almost nothing of interest.

In May and June 1977, while on a brief research trip to Leningrad at the very end of my stay, I became acquainted with the excellent group of historians of evolutionary biology working in the Sector on Evolutionary Theory and History of the Leningrad Branch of the Institute of the History of Science and Technology of the USSR Academy of Sciences, headed by Kiril M. Zavadskii. When I explained my difficulties, Zavadskii arranged for his student Eduard Kolchinskii—then a Party member—to act as an independent "consultant." Kolchinskii sat next to me in the archives, passing along file after file from the Filipchenko collection. It was there, sitting in the manuscript division of the Leningrad Public Library on a balmy spring day, that we came upon all of Dobzhansky's letters to Filipchenko.

With only days remaining in Russia, I arranged for copies of the correspondence to be exchanged between the Library of the American Philosophical Society (where Dobzhansky's archives, and therefore the Filipchenko letters, were kept) and the Leningrad Public Library, so that Dobzhansky's letters would become available. When I returned from Russia, the American Philosophical Society was (somewhat cautiously) willing, but the Leningrad Public Library demurred. Letters went back and forth over a three-year period, until all patience was exhausted and the matter languished. In a way the lapse was understandable. The history of Soviet genetics was still a sensitive topic in Brezhnev's Russia, for reasons I have recounted elsewhere ("The Soviet Nature-Nurture Debate," in *Science and the Soviet Social Order*, ed. Loren Graham, Cambridge: Harvard University Press, 1990, pp. 94–138), and Dobzhansky still labored under the stigma of the *nevozvrashchenets* ("nonreturner").

Only after a decade-long struggle was the exchange of letters effected, and then only after the dawn of Gorbachev's glasnost. In its early months (late fall 1986) I received a message, then a letter, from Kolchinskii suggesting that the exchange might now be feasible and that I should visit Leningrad at the earliest possible moment. At this point the International Research and Exchanges Board (IREX) came to the rescue. The American Council of Learned Societies and the (then) Soviet Academy of Sciences had established a Joint Commission on the Social Sciences and Humanities to promote joint Soviet-American scholarly work. Loren Graham headed one of its many subunits, the Subcommission on the History, Philosophy, and Social Studies of Science, and he agreed to include "Dobzhansky and the Evolutionary Synthesis: East and West" as its fifth research project, with me as the U.S. coordinator. At the joint planning meeting, the project was news to the Soviet cosponsors and they understandably demurred; only thanks to the insistent persuasion of Loren Graham and Wesley Fisher was the project agreed to. They are due special thanks; without their efforts it is doubtful that this volume would have been possible.

In spring 1987 I traveled to Leningrad on IREX funding. Ted Carter, Librarian of the American Philosophical Society, had provided me with copies of Filipchenko's letters to Dobzhansky on the off chance that an actual exchange might be effected. When I arrived in Leningrad, I renewed my acquaintance with Eduard Kolchinskii, Yasha Gall, Sergei Orlov, and A. B. Georgievskii, the last of whom had become director of the sector following Zavadskii's death. I also met three of their younger colleagues who would prove instrumental in subsequent events: Daniel Alexandrov (a biologist by training); Nikolai Krementsov (who had recently left the Pavlov institute); and the exarchivist most responsible for the resurgence of Soviet interest in Dobzhansky, Mikhail Konashev. Through means I still do not fully understand, it was Konashev who was able to see to it that the correspondence exchange took place. He also introduced me to Sergei Inge-Vechtomov, the dean of biology at Leningrad University and head of its Department of Genetics—the descendent of the department founded by Filipchenko where Dobzhansky had worked. To make the exchange possible, a plan was worked out for meetings and joint publications on Dobzhansky.

In late spring 1990 I received an invitation from my Leningrad colleagues to serve as cochairman of a meeting on Dobzhansky to be held in Leningrad in September of the same year. I was asked to arrange Western participation, and once again IREX came to the rescue, funding the travel of several American scholars who otherwise would not have been able to attend. Unfortunately, several invitees, including Ernst Mayr, Francisco Ayala, Richard Lewontin, Stephen Jay Gould, and Jeff Powell, had scheduling conflicts or other problems that prevented them from attending.

The International Symposium on Theodosius Dobzhansky was held in Leningrad 17–19 September 1990 and was hosted by the Leningrad Center of the USSR Academy of Sciences and the International Foundation for the History of Science in Leningrad. In addition to Sergei Inge-Vechtomov, my cochairman and our host, the organizing committee included Mikhail Konashev, Nikolai Krementsov, and Sergei Orlov. Among the speakers were those whose essays appear here. For the first time in many years, Dobzhansky—who had been regarded as an "unperson" in the USSR—was an object of respect and analysis in his original homeland.

In the course of the meeting, the exceptional quality and interest of several of the papers led many of us to believe that they would be of interest to a broad audience. A group of Americans and Russians met and discussed the options. It was decided to exclude from the prospective collection those technical scientific papers that did not centrally feature Dobzhansky and to exercise a strong editorial hand; instead of a volume devoted to conference proceedings, we sought one that would provide scholarly and interpretive studies of the life and legacy of one of the central figures in modern evolutionary biology. We are most gratified by the support and interest shown by Emily Wilkinson and the entire staff of Princeton University Press.

During my days at Mather Camp in early July 1974, Dobzhansky recounted for me his abortive attempts to visit Russia during the 1960s and early 1970s. He told me that although he never had any intention of staying for longer than a visit, he would have liked to see the old haunts of his youth, to visit family and old friends, to lend help where he could, and perhaps to be recognized at long last in the homeland that had disowned him. It was never to be. Had Dobzhansky lived to attend our meeting, he would surely have been gratified by the recognition he had so long awaited. With the ending of a half century of isolation, Russians gathered with citizens of the broader world to begin to reunite the threads that Dobzhansky's own life had woven together. I am grateful to have played a part in that beginning.

Mark B. Adams
Philadelphia
24 July 1993

ACKNOWLEDGMENTS

THIS VOLUME would not have been possible without the help and support of many people and organizations.

In the United States, special thanks are due to Ted Carter and the Library of the American Philosophical Society for their central role in the exchange of archival documents; to IREX, and especially to Wesley Fisher, for funding much of the travel associated with the project; and to the Joint Commission on the Humanities and Social Sciences of the American Council of Learned Societies and the Russian Academy of Sciences, and especially to the American coordinator of its Subcommission on the History, Philosophy, and Social Studies of Science, Loren Graham.

In Russia, special thanks are due to Sergei Inge-Vechtomov (St. Petersburg Scientific Center, Department of Genetics of St. Petersburg University) for sponsoring and supporting the meeting with such informality and grace; to the International Foundation for the History of Science, and especially to Nikolai Krementsov, Sergei Orlov, and other members of the organizing committee; to Nikolai Vorontsov and Yasha Gall for their wisdom and good advice; to Eduard Kolchinskii, who made possible the discovery and exchange of Dobzhansky's letters in the first place; and to Mikhail Konashev, whose single-mindedness and persistence are most responsible for the Russian rediscovery of Dobzhansky.

Work on this volume was partially supported by grants from the History and Philosophy of Science and the Science, Technology, and Society divisions of the National Science Foundation. Thanks go also to the University of Pennsylvania, its Research Foundation, and my colleagues in the Department of the History and Sociology of Science for their help and support. Finally, I am grateful to my colleagues for contributing the fruits of their research to this volume and especially to Sophie Dobzhansky Coe for sharing her family story with us all.

INTRODUCTION

Introduction:
Theodosius Dobzhansky in
Russia and America

Mark B. Adams

THEODOSIUS DOBZHANSKY (1900–1975) was one of the most important biologists of the twentieth century.

Consider his achievements. The central architect of the modern evolutionary synthesis during the interwar years, he integrated diverse biological specialties in his remarkably influential classic, *Genetics and the Origin of Species* (1937)—a book that reoriented the thinking of many biologists and whose subsequent editions constituted the evolving *locus classicus* of the new view. As a practicing scientist, he created population genetics in the United States, established it as the central evolutionary discipline, trained an international school in the subject, and created an ongoing series of articles, the "Genetics of Natural Populations" series, that formed the backbone of the field for twenty years. As a leading figure in American science, he wrote popular books and articles on social and philosophical issues, including race, equality, human evolution, religion, cosmology, Soviet science, and Lysenkoism. Almost twenty years after Dobzhansky's death, we still think about evolution the way he taught us to, and his legacy lives in the language we use: it was he who introduced such terms as *microevolution, macroevolution,* and *gene pool* into English.[1]

Yet in light of Dobzhansky's central role in twentieth-century genetics and evolutionary biology, he has not yet received the historical attention one might expect. Although recent years have seen autobiographies, biographies, or historical collections on most major biologists of the century, as well as most major architects of the evo-

---

[1] On his use of *microevolution* and *macroevolution*—terms he got from his teacher Filipchenko (1927), who first coined them—see Adams 1988, 1990b, and 1990g. On *gene pool* (an adaptation of Serebrovskii's 1926 term *genofond,* "gene fund"), see Adams 1979 and 1990e.

lutionary synthesis (e.g., R. A. Fisher, J. B. S. Haldane, and Sewall Wright), nothing comparable has appeared on Dobzhansky. In recent decades a vigorous historical literature on twentieth-century biology and the evolutionary synthesis has burgeoned, and in much of it Dobzhansky has emerged as a major—often *the* major—figure.[2] But thus far, works centering on Dobzhansky are limited to a few historical essays, two volumes of reprints, and a few festschrifts.[3]

In one sense, the relative historical neglect of Dobzhansky is understandable. Both his 1937 classic and his population genetics seemed to burst on the scene in the late 1930s out of nowhere; in his various technical publications from the Morgan lab in the preceding decade, there was hardly a hint of what was to come. The search for the origins of Dobzhansky's scientific theory and practice, however, was stymied by a lack of information about his formative years, which he spent in Kiev and Leningrad.[4] He left the Soviet Union for New York in 1927 to work in the laboratory of T. H. Morgan, the founder of modern genetics, and never returned to Russia.

This volume is a product of the rediscovery of Dobzhansky in his original homeland. That rediscovery was a long time in the making.

[2] See, e.g., Smocovitis 1992.

[3] Dobzhansky's unpublished oral memoirs, based on interviews taped in 1962 by Barbara Land for Columbia University's Oral History Project (Dobzhansky 1962–1963), remain the major source on his life; they were conducted in English and do not deal with his scientific work in any detail, nor does the interviewer's popular children's book based on them (Land 1973). A compendium of Dobzhansky's travel letters, assembled by Glass (1980), also includes a long passage taken verbatim from a copy of the memoirs I provided him. On occasion Dobzhansky himself reworked some of the material into elegant historical articles (1980a, 1980b). Like most memoirs, however, they were dictated long after the events they describe, and they leave out much that Dobzhansky apparently regarded as being of little interest to American readers. His classic papers that form the "Genetics of Natural Populations" series have been reprinted (Lewontin et al. 1981). The best secondary historical source in English on Dobzhansky's Russian years is Provine's introductory essay in that volume, based largely on Dobzhansky's memoirs and his publications in Western languages. The principal biography of Dobzhansky is that written by his former student Francisco Ayala (1990); here, too, the very brief account of Dobzhansky's Russian years is taken largely from the oral memoirs, and there is no discussion of the Russian roots of Dobzhansky's thought. Other recent articles on Dobzhansky (e.g., Beatty 1987, Lewontin 1987, Crow 1987) have dealt principally with his role in technical biological debates in the postwar years.

[4] St. Petersburg was renamed Petrograd (1914–1924), then Leningrad (1924–1991), then St. Petersburg (1991–). During those periods, other institutions' names employing the city in their titles changed accordingly (e.g., St. Petersburg University). In general, throughout this volume references to the city and to the names that depend on it are given in the form appropriate to the period being discussed.

Following the advent of Stalinism in the 1930s, Dobzhansky was vilified in the USSR as a "traitor," a "nonreturner" (*nevozvrashchenets*), an "enemy of the People" (*vrag naroda*). As an "unperson," he could not be cited or mentioned in publications, except derisively. After World War II, Dobzhansky was often featured prominently in the public press during Lysenko's heyday as a "fly-lover and man-hater," a tool of machinating capitalists and American imperialism. Even after 1965, with genetics reestablished in Russia and Lysenko dethroned, the earlier invective still cast a shadow on Dobzhansky's name. As recently as the early 1970s, Dobzhansky was refused a Soviet visa to return to Russia for a last visit. Although able to mention Dobzhansky (at least as an American scientist), Soviet publications still tended to treat him as a figure of secondary importance. It took a full decade after Dobzhansky's death, and the advent of Gorbachev's policy of glasnost in 1986, for Soviet historians to begin, very tentatively, to acknowledge Dobzhansky as one of their own. The acknowledgment went public only in March 1990.[5] The definitive recognition came six months later with the International Symposium on Theodosius Dobzhansky. Held in Leningrad the week of 17 September 1990, it celebrated Dobzhansky's scientific legacy in the very settings where he himself had worked seventy years before. That meeting first aired much of the work from which this book was shaped.

Thanks to the recent "reopening" of Russia, this volume is able to explore in depth one of the central unanswered questions of Dobzhansky's legacy: the impact of his Russian background on his subsequent contributions to evolutionary biology. That Dobzhansky's first twenty-seven years in Kiev and Leningrad should profoundly

[5] See the section in the March 1990 issue of *Priroda* entitled "Nash Dobrzhanskii" [Our Dobzhansky]. It opens with a portrait of the young Dobzhansky in Leningrad, with the following caption: "The 90th anniversary of the birth of Feodosii Grigor'evich Dobrzhanskii, one of the leading biologists of our day, geneticist and evolutionist, is being widely observed by our American colleagues, who consider Dobzhansky their own American scientist. Indeed, in all biographical references one reads that Dobzhansky was an American geneticist, one of the founders of experimental population genetics and of the synthetic theory of evolution. But this isn't exactly so. Placed by fate in one of the best genetics laboratories in the world, headed by the creator of the chromosomal theory of heredity T. H. Morgan, and drawing upon the ideas of his teachers—the great Russian naturalists V. I. Vernadskii, N. I. Vavilov, S. S. Chetverikov, and Iu. Filipchenko—he was able to arrive at the synthetic theory of evolution by uniting the experiences of Russian and American biological schools" (*Priroda*, 1990, no. 3, p. 78).

affect his subsequent work and thought is hardly remarkable. Yet until now, the Russian archives and sources that would clarify his Russian years have been largely inaccessible to Western historians. Thus, perforce, Western science historians have largely omitted discussion of the possible Russian origins of Dobzhansky's ideas and approaches, sometimes treating him as a scientist whose work was purely a product of the American scene.

The omission of Dobzhansky's Russian dimension has contributed, in turn, to a curious feature of the literature on the evolutionary synthesis of the interwar years: although most of that literature has declared the synthesis to have been a broadly international event, it has portrayed that synthesis in almost exclusively Anglo-American terms.[6] It is hoped, then, that this volume—the first book devoted to an analysis of the historical, scientific, and cultural dimension of Dobzhansky's life and thought—will not only provide much new information on a central figure in twentieth-century biology but, given its rich analysis of Dobzhansky's Russian background, that it will also help us understand the international dimensions of the development of population genetics and the evolutionary synthesis as a whole.

All of the essays in this volume were written for it and contain new work. Many of the essays draw upon manuscript sources never before available, ranging from Dobzhansky's family album through previously inaccessible archives in St. Petersburg and Kiev and including unpublished correspondence dating from 1918 through the end of his life. Even the best standard biographical accounts of Dobzhansky's life have left unanswered many questions of real importance—for example, the nature of his work in Kiev and Leningrad, the interplay of Russian and American traditions in his classic works, his positions on the social and political issues that vexed the United States when he was one of its leading scientific figures, and the nature and origin of his deeper philosophical and religious concerns. In this volume, Russian and American archives have been mined to explore these issues in detail.

---

[6] The great exception, of course, is Mayr and Provine 1980, which devotes a section to analyzing the evolutionary synthesis in different countries. Nonetheless, it is an exception that proves the rule: the work emphasizes events in the United States and Britain, whereas several of the essays on other countries (notably France and Germany) seek to understand why they were so far behind Anglo-American developments.

The volume opens with a family portrait of Dobzhansky sketched by his daughter, Sophia Dobzhansky Coe. Based on her own reminiscences as well as on her father's papers and photographs from the family album, her account ranges from the family's early roots in Poland to charming stories of T. H. Morgan, Pasadena, life in New York, international travels, and the final years of her father's life. Based on her Leningrad presentation (which she delivered, to the astonishment of all present, in almost simultaneous Russian and English), her story provides a most fitting opening. Following this account, the volume is divided into four parts: Russian Roots; The Morgan Lab; The Scientific Legacy; and Dobzhansky's Worldview.

Russian Roots

The man who became "Doby" began life as a bright young Russian naturalist named Feodosii Grigor'evich Dobrzhanskii. What we have known about the young Dobzhansky has until now been limited to what he himself told us in his oral memoirs and to a few other sources.[7] In this section our understanding is greatly enriched by the important contributions of Russian historians of science. Bringing to their work a familiarity with both Russian and Western published sources, as well as hitherto inaccessible Russian archival materials, they provide us with much new information not only about Dobzhansky's early life but also about the roots of some of his seminal ideas. Nikolai Krementsov's article documents Dobzhansky's involvement in the Russian entomological community, arguing that its theory, practice, and structure had a profound effect on Dobzhansky's subsequent work and were the source not only for his views on speciation but also for the biological species concept for which he became well known. Daniel Alexandrov also highlights the entomological dimension of Dobzhansky's work as part of a broader comparison of Filipchenko and Dobzhansky as zoologists, a comparison that explores for the first time Dobzhansky's Russian publications. Finally, Mikhail Konashev presents the results of several years of archival discovery, providing us with new information about Dobzhansky's family background, his work in Kiev, and Filipchenko's Leningrad school.

---

[7] See Note 3 above; Adams 1968, 1979, 1988, 1990b, and 1990g; and Provine 1981.

The Morgan Lab

Less than a decade after his arrival in the United States in December 1927, Dobzhansky published his evolutionary classic, *Genetics and the Origin of Species*, and launched American population genetics. The essays in this section explore Dobzhansky's achievements during his time in the Morgan lab in terms of the interaction of his Russian background with the American setting. Garland Allen, drawing on an unpublished 1967 interview, discusses Dobzhansky's role in the Morgan group in uniting field and laboratory practice. William Provine mines many unpublished sources to explore how *Genetics and the Origin of Species* came to be written in 1936 and early 1937. Robert Kohler analyzes Dobzhansky's population genetics as a new "system of scientific production" (based on "wild" *Drosophila pseudoobscura*) in the context of the "moral economy" of the Morgan group. Finally, Richard Burian compares the various editions of Dobzhansky's 1937 classic in order to explore the roots in Russia (and Filipchenko) of his treatment of evolutionary dynamics and the problem of macroevolution.

The Scientific Legacy

In this section, three biologists discuss Dobzhansky's scientific legacy. Scott Gilbert analyzes Dobzhansky's treatment of embryological issues, exploring his preference for the views of Schmalhausen over those of Waddington. Bruce Wallace draws on a quarter century of interaction with Dobzhansky to analyze his evolving attitudes toward the issue of genetic coadaptation. Finally, Charles Taylor, one of Dobzhansky's last students, recalls their exchanges on the "larger questions" of evolution and the way in which Taylor's own work on computer modeling is beginning to address them.

Dobzhansky's Worldview

The final section of the volume explores Dobzhansky's social, political, philosophical, and religious views. In the opening essay, Dobzhansky's student Costas Krimbas surveys Dobzhansky's broad evolutionary worldview, tracing the relationship of his scientific work to his broader concerns with human heredity, human evolution, meaning, destiny, and religion. John Beatty treats Dobzhansky as a "biologist of democracy," discussing his views on the moral and political significance of genetic variation and assessing the influence of his American setting (Columbia in the late 1930s and 1940s) on his political philosophy. Diane Paul chronicles Dobzhansky's books and articles on human heredity to assess his position in the ongoing American "nature-nurture" debate. Finally, Michael Ruse analyzes Dobzhansky's view of progress in the broadest sense, exploring the apparent contradiction between his materialist evolutionary views and his fascination with Teilhard de Chardin to find an underlying philosophical unity in Dobzhansky's worldview.

The volume thus reflects the full breadth of Dobzhansky's life and thought. As such, it should be of interest not only to students of evolutionary biology, genetics, American science, international scientific relations, and the history of Russian science but also to those interested more broadly in science and culture, the interwar "intellectual migration," the politics of knowledge, scientific practice, and the relationship between biological, philosophical, and social thought.

References

Adams, Mark B. 1968. "The Founding of Population Genetics: Contributions of the Chetverikov School, 1924–1934." *Journal of the History of Biology*, vol. 1, no. 1, pp. 23–39.
———. 1979. "From 'Gene Fund' to 'Gene Pool': On the Evolution of Evolutionary Language." In *Studies in History of Biology*, vol. 3, ed. William Coleman and Camille Limoges (Baltimore: The Johns Hopkins University Press), pp. 241–85.
———. 1980. "Sergei Chetverikov, the Kol'tsov Institute, and the Evolutionary Synthesis." In *The Evolutionary Synthesis: Perspectives on the Unification of Biology*, ed. Ernst Mayr and William Provine (Cambridge: Harvard University Press), pp. 242–78.
———. 1988. "A Missing Link in the Evolutionary Synthesis" (essay review of Dobzhansky's edition of I. I. Schmalhausen, *Factors of Evolution*). *Isis*, vol. 79 (June), pp. 281–84.
———. 1990a. "Chetverikov, Sergei Sergeevich." In *Dictionary of Scientific Biography*, vol. 17, supp. 2, ed. Frederic L. Holmes (New York: Charles Scribner's Sons), pp. 155–65.
———. 1990b. "Filipchenko, Iurii Aleksandrovich." In *Dictionary of Scientific Biography*, vol. 17, supp. 2, ed. Frederic L. Holmes (New York: Charles Scribner's Sons), pp. 297–303.

Adams, Mark B. 1990c. "Karpechenko, Georgii Dmitrievich." In *Dictionary of Scientific Biography*, vol. 17, supp. 2, ed. Frederic L. Holmes (New York: Charles Scribner's Sons), pp. 460–64.

———. 1990d. "Levitskii, Grigorii Andreevich." In *Dictionary of Scientific Biography*, vol. 18, supp. 2, ed. Frederic L. Holmes (New York: Charles Scribner's Sons), pp. 549–53.

———. 1990e. "Serebrovskii, Aleksandr Sergeevich." In *Dictionary of Scientific Biography*, vol. 18, supp. 2, ed. Frederic L. Holmes (New York: Charles Scribner's Sons), pp. 803–811.

———. 1990f. "The Soviet Nature-Nurture Debate." In *Science and the Soviet Social Order*, ed. Loren Graham (Cambridge: Harvard University Press), pp. 94–138.

———. 1990g. "La génétique des populations était-elle une génétique évolutive?" In *Histoire de la Génétique: Pratiques, Techniques et Théories*, ed. Jean-Louis Fischer and W. H. Schneider (Paris: A.R.P.E.M.), pp. 153–71.

———. 1991. "Through the Looking Glass: The Evolution of Soviet Darwinism." In *New Perspectives on Evolution*, ed. L. Warren and H. Koprowski (New York: Wiley-Liss), pp. 37–63.

Ayala, Francisco J. 1990. "Dobzhansky, Theodosius." In *Dictionary of Scientific Biography*, vol. 17, supp. 2, ed. Frederic L. Holmes (New York: Charles Scribner's Sons), pp. 233–42.

Beatty, John. 1987. "Weighing the Risks: Stalemate in the Classical/Balance Controversy." *Journal of the History of Biology*, vol. 20, no. 3 (fall), pp. 289–319.

Crow, James F. 1987. "Muller, Dobzhansky, and Overdominance." *Journal of the History of Biology*, vol. 20, no. 3 (fall), pp. 351–80.

Dobzhansky, Theodosius. 1937. *Genetics and the Origin of Species*. New York: Columbia University Press.

———. 1962–1963. "The Reminiscences of Theodosius Dobzhansky." Typed transcript. 2 parts. Oral History Research Office, Columbia University, New York.

———. 1980a. "The Birth of the Genetic Theory of Evolution in the Soviet Union in the 1920s." In *The Evolutionary Synthesis: Perspectives on the Unification of Biology*, ed. Ernst Mayr and William Provine (Cambridge: Harvard University Press), pp. 229–42.

———. 1980b. "Morgan and His School in the 1930's." In *The Evolutionary Synthesis: Perspectives on the Unification of Biology*, ed. Ernst Mayr and William Provine (Cambridge: Harvard University Press), pp. 445–552.

Filipchenko, Iu. A. [Philiptschenko, Jur.] 1927. *Variabilität und Variation*. Berlin: Gebrüder Borntraeger.

Gall, Ia. M., and M. B. Konashev. 1990. "Klassik" [Classic]. *Priroda*, no. 3, pp. 79–87.

Glass, Bentley, ed. 1980. *The Roving Naturalist: Travel Letters of Theodosius Dobzhansky.* Philadelphia: American Philosophical Society.

Hecht, Max K., and William C. Steere, eds. 1970. *Essays in Evolution and Genetics in Honor of Theodosius Dobzhansky.* New York: Columbia University Press.

Kohler, Robert E. 1991. "Drosophila and Evolutionary Genetics: The Moral Economy of Scientific Practice." *History of Science,* vol. 29, pp. 335–75.

Konashev, Mikhail B. 1991. "Ob odnoi nauchnoi komandirovke, okazavsheisia bessrochnoi" [The postdoc that never ended]. In *Repressirovannaia nauka* [Science under repression], ed. M. G. Iaroshevskii (Leningrad: Nauka), pp. 240–63.

Land, Barbara. 1973. *Evolution of a Scientist: The Two Worlds of Theodosius Dobzhansky.* New York: Crowell.

Lewontin, R. C. 1987. "Polymorphism and Heterosis: Old Wine in New Bottles and Vice Versa." *Journal of the History of Biology,* vol. 20, no. 3 (fall), pp. 337–49.

Lewontin, R. C., J. A. Moore, W. B. Provine, and B. Wallace, eds. 1981. *Dobzhansky's Genetics of Natural Populations I–XLIII.* New York: Columbia University Press.

Mayr, Ernst, and William Provine, eds. 1980. *The Evolutionary Synthesis: Perspectives on the Unification of Biology.* Cambridge: Harvard University Press.

Provine, William B. 1981. "Origins of the 'Genetics of Natural Populations' Series." In *Dobzhansky's Genetics of Natural Populations I–XLIII,* ed. R. C. Lewontin, J. A. Moore, W. B. Provine, and B. Wallace (New York: Columbia University Press), pp. 1–83.

Smocovitis, V. B. 1992. "Unifying Biology: The Evolutionary Synthesis and Evolutionary Biology." *Journal of the History of Biology,* vol. 25, no. 1 (spring), pp. 1–65.

Sorokina, M. Iu. 1990. "Dal'nii put' k bol'shomu budushchemu: Iz perepiski F. G. Dobrzhanskogo s V. I. Vernadskim" [The long path into the big future: from the correspondence of F. G. Dobrzhanskii and V. I. Vernadskii]. *Priroda,* no. 3, pp. 88–96.

Theodosius Dobzhansky:
A Family Story

Sophia Dobzhansky Coe

THEODOSIUS DOBZHANSKY, my father, was, by the Gregorian calendar, born on 25 January 1900 in Nemirov, a town southwest of Kiev. By the Julian calendar, in use in the Russia of the time, it was 12 January. He often said that had he been born twelve days earlier, that is to say in 1899 by the Julian reckoning, he would have been subject to the draft for World War 1 and his history would have almost certainly been a different and a shorter one.

Nowhere can I find the date of his parents' marriage. This has a bearing on his name of Theodosius, which is as unusual in Russia as it is anywhere else. Even the standard list of given names in the back of the dictionary does not give it. I believe it means "gift of God" in Greek, which would be very suitable given the circumstances. My father's parents were childless for quite a while after their marriage and tried to remedy their condition by prayer and pilgrimage. One of their pilgrimages led them to the shrine of St. Theodosius of Chernigov, a town roughly as far to the northeast of Kiev as Nemirov was to the southwest. Whether the shrine of St. Theodosius of Chernigov was known for making the sterile fertile, or whether some other reason led them to the spot, the prayers, and the vow to name the infant for the saint, were effective, and the child was born during the second week of what was, technically, not the first year of the new century but the last year of the old.

The father, Grigory Karlovich Doberzhansky, was a mathematics teacher in the Nemirov high school and the youngest son of a large family. His father, Karl Kazimirovich Doberzhansky, had been a Polish landowner in Kiev Province, but his lands were confiscated when he took part in the Polish uprising of 1863. He had married, without her father's permission, a Countess Tyshkevich, and the family romance would have it that the confiscated lands fell into the hands of the angry father, who left his daughter and her five children in poverty while Karl Kazimirovich spent twenty years of exile in Kargopol,

Olonetzk Province, in northern Russia. The family was originally Catholic but converted to Orthodoxy when Grigory Karlovich was a child. The only traces preserved of this Polish connection are a very Italianate oil on the canvas family icon of the Virgin and Child and the tattered remnants of a Polish cookbook.

The mother of the infant, Sophia Vasilievna Voinarsky, was three years younger than her husband and, like him, born in Kiev Province. She was a collateral relation of Dostoevsky, being the granddaughter of the author's father's sister. This was always a source of great pride to my father and provided a golden opportunity for those who like to hunt for things like writing ability among the branches of family trees. Sophia Vasilievna gave her son much of his earliest education after it was discovered that because of his German nursemaid he spoke better German than Russian.

In 1910 Grigory Karlovich had an accident and was no longer able to teach. The family moved to Kiev, and he died there in 1918. Sophia Vasilievna survived two years and died in May 1920. My father always believed that she died of a heart attack suffered while trying to eat a particularly hard and dry chunk of bread he had found for her, there being a famine in Kiev at the time.

There are a few anecdotes from my father's school years in Kiev that hint that there was a scientist in the making. He always remembered the action, if not the name, of the teacher who first allowed him free access to the school microscope and finally permitted him to take the microscope home. There were two summer trips to the Caucasus, the first a school trip in 1914 and the second, independently undertaken, in 1916. This second time my father went with a friend, Vadim Alexandrovsky, and the story was that each boy persuaded his mother that they were going to the Caucasus with the other boy and his mother. The deception was a success, until the two ladies, who knew each other by sight, met one day in Kiev. Vadim's mother went out to look for them, but finding them capable of taking care of themselves, gave them some cash and returned to Kiev.

It was during this period that a generalized interest in collecting insects crystalized into specialization in Coccinelids, the ladybugs. I remember being told that at one point in his youth my father's greatest ambition was to capture a specimen of every species available in the Kiev area. He retained his affection for these creatures all his life, being exceedingly pleased to see many years later in our vegetable garden in northwestern Massachusetts a ladybug that was familiar to him from his youth.

Fig. 1. Dobzhansky's mother,
Sophia Vasilievna Voinarsky.

Fig. 2. Dobzhansky's father,
Grigory Karlovich Doberzhansky.

During this period of deepening disorder my father was, owing to the fortunate accident of his birth date, able to avoid military service, finish high school, and begin university. The upheaval brought opportunities as well. During the summer of 1919 two distinguished scientists, V. Vernadsky and S. Kushakevich, took refuge in what had recently become the Dnieper Biological Station but had previously been the house of the director of the Staroselie forestry district. The family of the forester, Peter Alexandrovich Sivertsev, was allowed to keep three rooms of the house, and the other six became the field station. My father brought the mail and food from Kiev and became acquainted with the forester's daughter, Natalia Petrovna Sivertseva, who eventually became my father's wife and my mother. (Today the entire site of all these meetings is underwater, having been drowned by a hydroelectric dam.) Other ties were formed at the same time. Vernadsky's daughter, Nina, accompanied her father and became a lifelong friend of my mother. My mother died in February 1969 during a visit she was making to Nina Vernadsky Toll in Middletown, Connecticut.

My mother was also a biologist by training. She and my father were married on 8 August 1924 in the cave church of the Kiev-Pechersk Lavra, a monastic community organized in the eleventh century by another St. Theodosius, this one a monk, and having nothing in common but the name, so far as I know, with St. Theodosius of Chernigov.

Sometime during the early twenties my father began to learn English, to be able to read the new material coming from the West. Although he spoke English with a strong and distinctive accent, his written English was superb and could only be distinguished from that of a native by an inability to ever completely master the bewildering English usage of *a* and *the*.

By the time of my parents' marriage, my father had transferred his scientific activities from Kiev to Leningrad, where he arrived early in 1924. On his departure the Kiev group gave him a signed photograph inscribed "to dear ThD from his colleagues."

Winters were spent in a basement laboratory somewhere along the Neva on Universitetskaya Naberezhnaya. I remember my father's tales of one of the last great Neva floods when I was reading him Pushkin's *Bronze Horseman,* one of the many Russian classics that I read him over the years. Summers were in part spent in laboratories on the outskirts of Leningrad, some inserted into the great suburban

Fig. 3. The photo given to Dobzhansky in Kiev as a going-away present, inscribed "to dear ThD from his colleagues." *Top row, left to right*: Sergei Ivanov, Longin Kosakovsky, Iulii Kerkis (who shortly thereafter followed Dobzhansky to Leningrad). *Bottom row, left to right*: M. M. Levit, Dobzhansky, A. G. Lebedev, N. S. Greze, and G. I. Shpet.

palaces of the prerevolutionary nobility. There must have been quite a group from Kiev because besides my father's student Iulii Kerkis, my mother's brother Alexander Petrovich Sivertsev, a chemist, lived there and found a bride in Anna Andreevna, who figures in some of the group pictures of the Summer Institute of Biology.

Philipchenko was the director of the genetics laboratory, and he was responsible for my father's three summer trips to Central Asia to study domesticated animals. They provided for another group of family anecdotes. One had to do with sitting in a nomad tent and, as the honored guest, being offered the sheep's eyes. Fortunately for the honored guest, it was the most exquisite politeness to say that the honor was too great and to return the eyes to the host. There was a story of bravado having to do with a horse, presumably unbroken, which gave my father a rough ride over the steppe but did not succeed in dislodging him, thus impressing the locals. At one point the expedition camped by a lake, and a discussion arose about a distant flock of birds. My father seized a gun, and in what was literally a long shot, killed his prey. He always said it was the last time he ever shot

Fig. 4. Dobzhansky on expedition to Kazakhstan in the mid-1920s, shortly before leaving for New York.

anything. For all I know, it was also the first time. My mother, already the endlessly hospitable and welcoming helper, remembered with a shudder the time one of the important men from Central Asia came to visit them in Leningrad and she prepared a meal that featured ham and red wine, neither of which the Muslim bigwig was permitted to consume.

It was Philipchenko who suggested the Rockefeller grant for one year of study in the United States, which brought my parents to New York in the last days of 1927. They were appalled by the griminess of T. H. Morgan's laboratory in Columbia's Schermerhorn Hall and especially remembered the cockroaches in the desk drawers. T. H. Morgan became far more than just a scientific mentor to my parents. He taught them English in his rather whimsical manner, giving pithy bits of advice such as "You can say 'I smell,' but you cannot say 'You smell'." He was also their introduction to what can be called American culture in the anthropological sense. My father was always rather distressed at Morgan's lack of interest in philosophical questions, but he prized the memory of being prodded to find his own scientific problems rather than relying on directions from the boss.

Morgan was also a mentor for the younger generation of Dobzhanskys. One of the few things I recall about his house in Pasadena, aside from the avocado tree and the question as to when, if ever, it would bear fruit, was the firm advice given to me at one meal, that salad could only be eaten with the fingers if there was no dressing on it—an admonition the young savage never forgot.

Another important figure during this formative period was Miss Wallace, always so referred to by my parents, although her given name was Edith. Nominally she was Morgan's secretary, but her great talent lay in painting meticulous portraits of *Drosophila*. Magnified until the image is almost a foot high, and with the thickness, length, and angle of every bristle correct, they are extraordinary pieces of natural history illustration. Miss Wallace was another instructor in American mores and language.

The Morgan laboratory moved to the California Institute of Technology in Pasadena in 1928. Given the Russian romanticization of warm climates, it must have been amazing for my parents to move to a place where palm trees lined the streets and not-too-distant orange groves perfumed the air. The house in Pasadena that I remember had three orange trees, two peach trees, an apricot tree, and an enormous fig tree, as well as black widow spiders in the garden shed, which I was warned never to visit. The thought of eliminating the spiders never seemed to enter anyone's head. For a Russian, the pleasure of climbing about in the orange trees in his garden to knock down fruit for the breakfast juice must have been enormous.

The early thirties were not all idyllic. There were two related problems, the first being whether or not to return to the Soviet Union. I have recently seen a copy of a letter from Vavilov, sent in 1930, urging return with every sort of argument from patriotism to quoting Morgan as feeling that he should not impede return. It must have been an extraordinarily difficult decision to make, but it was made simpler by the fact that, as time passed, it became increasingly obvious that all was not going well in the Soviet Union. A letter arrived from a respected colleague, written apparently in response to a letter hinting at an inclination to come back, with "and I thought you were a clever man" written on it somewhere in very small writing. I saw this letter when I was sorting my father's papers after his death and directed it to the American Philosophical Society Library with the rest of his papers, but no one there seems to have seen the all-important message, and the identity of the sender escapes me.

Fig. 5. The Morgan group, Pasadena, 1930.
Top row, left to right: M. Roades, R. L. Biddle, W. A. Hetherington,
Goulding (a visitor). *Second row, left to right:* A. D. Zuluette (a visitor),
Dobzhansky, G. D. Karpechenko (a visitor), Karl Lindegren, Calvin Bridges,
A. Tyler, Dr. Wynegarden, Jack Schultz. *Bottom row, left to right:*
A. H. Sturtevant, T. H. Morgan, Karl Belar, Henry Borsook.

The second dilemma was no simpler. The U. S. Immigrations Service, exclusionist and racist as it was at the time, inflicted endless nightmares. At one point there was an agonizing trip to Vancouver to fulfill some sort of bureaucratic requirement, with a hinted suicide pact if permission to reenter the United States was denied. But even at this darkest moment the scientist was at work, because it was on this trip that my parents met Michael Lerner, who became a geneticist and a lifelong friend.

But these years of difficult decisions held pleasures as well. The purchase of a Model A Ford led to the exploration of California and neighboring states on camping trips, usually undertaken with a group of friends. It must have been this car in which my father took Vavilov to Sequoia National Park during his visit in 1931. There were also trips to Alaska, Mexico, and Guatemala. Eventually the trips became collecting trips for wild *Drosophila*, but before that there was a brief flirtation with *Pogonomyrmex* ants as experimental material,

Fig. 6. Dobzhansky and his wife, Natalia (Natasha), with guest
G. D. Karpechenko in front of the Dobzhanskys' house in Pasadena, 1930.
The photo was taken by Nikolai Vavilov, who was visiting.

which came to nothing. Closer to home there was always horseback riding. In an indirect way this led to the writing of *Genetics and the Origin of Species*. The horse was said to be blind in one eye, and it smashed my father's knee against a stone post. The kneecap was shattered into fourteen pieces and required an operation to insert a ring to hold the fragments together. The anesthetist had difficulties getting the ether-using *Drosophila* geneticist to sleep and suggested that he was drunk, this apparently inhibiting the action of the anesthetic. The operation called for a long recovery, and my mother rented a hospital bed. It is one of my earliest memories, seeing my father in what seemed a very high bed indeed, writing away on what was to become his first large book.

Another early memory is of my parents coming back to the house together on what seemed to be an unusually brilliant day. Their mood was equally sunny; in fact they were glowing. The long bureaucratic agony was over, and on that day, 8 October 1937, they had finally become naturalized American citizens.

This solidified my parents' ties to the United States, but it weakened their ties with the Soviet Union. In January 1938 they received a letter from my mother's brother, Alexander Petrovich, who lived

Fig. 7. From the family album: at the beach,
Natalia and Theodosius Dobzhansky with daughter Sophia.

with his wife and daughter in the same apartment in Leningrad that my parents had left in 1927, saying that they were deeply distressed by the decision of my parents to stay in the United States and that henceforth all communication with their sister and mother must cease. My maternal grandmother, Sophia Pavlovna Sivertseva, had joined us in 1935 after the death of her husband, the forester. Despite this open break, Alexander Petrovich was considered a suspect person for the rest of his life, and the family endured grave difficulties because of their connection to the "nonreturners" overseas.

In 1940 our family moved to New York City and my father to a professorship at Columbia University. The retirement of Morgan was one reason for this, perhaps the deterioration of relations with Sturtevant was another, although the fact that their daughter Harriet stayed with us on her way to get married in England after World War II shows that the two families did not entirely lose touch. A large factor was the provincialism of Pasadena and Caltech. There was at that time what might be called a nativistic movement, with Americans, especially academics, resenting the arrival of European scholars. Domestic snobbery existed as well, and there were obligatory gruesome soirées with the trustees of Caltech, when the faculty had to don black tie and be snubbed.

New York gave my father a far greater opportunity to enjoy his very conservative taste in music and art. Beethoven was to him the be-all and end-all of composers, and there was constant grumbling at the penchant of the Philharmonic for scheduling unacceptable modern pieces at the beginning of the performance, so as to make them unavoidable. If foreign visitors wished to see the Museum of Modern Art, they were taken there, but there were plenty of comments afterward about the absurdity of the objects on the walls. I never remember my father going to the theater with my mother and myself. The American Museum of Natural History was another matter altogether, and trips to see the exhibits, and Ernst Mayr, were frequent events.

Conservative in art as he may have been, my father was not so in politics. Roosevelt was his idol, and to the end of my father's days, when *liberal* was becoming the bad word it is now in some circles, he would proudly announce that he was a liberal. I do not think he ever voted for a candidate who was not a Democrat in a national election.

The pleasures of camping trips and excursions had to be foregone in the New York area because it was considered that these activities could only take place west of the one hundredth meridian, that is to say near the eastern foothills of the Rockies. My father's interest in the scenery of the eastern United States was confined to train trips to Cold Spring Harbor to visit the Demerecs.

Despite the war there were still summer projects in the San Jacinto mountains of southern California. This was the period of release and recapture experiments, when mutants marked by conspicuous red eyes were released in the middle of a cross-shaped pattern of traps, consisting of small cups of fermenting mashed bananas on wire stands. It was no small achievement to obtain bananas during the war, and sometimes the flies had to be content with a fermenting mass of dried banana flakes. Given the jitters of the period, it is not surprising that the presence of a person with a strong foreign accent doing such peculiar things was reported to the FBI. They came out and questioned my father, and he tried to explain the scientific thinking behind his activities. All his years of uncertain immigration status, plus his distrust of authorities rooted in horror stories coming out of the Soviet Union, combined to make this interview intensely distressing to him, and he envisioned the collapse of the research plans for that summer, if not something worse. Fortunately the FBI investigator, baffled by his initiation into the theory of genetics, could think of nothing better to do than ask the advice of a young nature counselor at a nearby Boy Scout camp. The counselor mulled it over, decided that it made sense to him, persuaded the FBI agent, and my father was off the hook. It was not until many years later, shortly before my father's death, when we were retelling this story over the dinner table, that the identity of my father's savior was disclosed. It was another guest at the table, Charles Remington, who had in his youth saved my father from the unwanted attention of the FBI and who had gone on to make entomology his life work.

World War II also provided an opportunity for travel in other directions. In 1943 the U. S. Government sent my father to Brazil as part of the Good Neighbor Policy. The Brazil connection continued for years and gave my father a chance to indulge in his love for the tropics. He often referred to passages in Charles Darwin's *Voyage of the Beagle* as his intellectual precedent for this delight in the abundance of tropical life, so different from the mindless fear felt by many people. Eventually he managed to travel to, and collect in, almost every state and territory, often on military planes, some of which were quite old and battered, although at that time they were the only means of transport.

One trip sticks in my mind, when I accompanied my father on a collecting foray to the state of Goias in the center of Brazil, not far from where the capital of Brazil, Brasília, now stands. The control panel of the plane included many gaping holes, some with loose ends of wires protruding. The course was plotted with a string on the map before we took off. Despite all this we arrived safely at our destination and collected our samples. My father sorted them, using the binocular microscope that traveled with us in a green metal ammunition case. I recorded the count of species using a technique he taught me, first making four dots on the corners of an imaginary square and then connecting the dots with lines, so that the final symbol, a box with a cross in it, stood for ten specimens, a graphic device I have never encountered elsewhere. This trip took place during the election between Harry Truman and Thomas Dewey, and my father naturally asked the military radio operator if he could tell us the results of the voting. He finally had to restrain his curiosity until we reached a larger town on our return journey, because all the soldier could tell him was that either Truman or Dewey had won!

Eventually the University of São Paulo gave my father an honorary degree for his services. Although in his later years he was given many honorary degrees, he always remembered this one with fondness because of the elaborate costume involved. It included a fluted cylindrical hat topped with a crystal sphere and a tunic of sky blue velvet to go over the black academic gown. He delighted in retelling the story of a newspaper reporter who, seeing someone else in the same rig, announced that he looked like the pope's mistress.

The Brazilians were a large and vocal contingent among the innumerable students and professors who passed through the Columbia laboratory and the nearby apartment at 39 Claremont Avenue. For a while coffee was made there in the Brazilian fashion, which involved a flannel bag full of coffee grounds suspended in the coffee pot, and the sugar consumption rose to such an extent that my mother persuaded them to get sugar ration cards, lest the family sugar supply be exhausted by the cafézinhos.

Other visitors came with other demands. J. B. S. Haldane arrived in the late forties, when civilian clothes were scarce goods indeed, and announced to the press that he would judge the United States by its capacity to supply him with the outsize shirts he wore. Needless to say it was my mother who represented the United States in this matter and, knowing that Alfred Mirsky had a brother-in-law who was a shirt manufacturer, she made a few phone calls and obtained the shirts. She said afterward that she did not really think Haldane needed a size that special, but Haldane was always recognized as a singular case.

Then there was the Chilean geneticist who performed the not uncommon trick of taking the wrong subway, so that he found himself at 116th Street and Lexington Avenue in Harlem, rather than at 116th Street and Broadway, near Columbia. By the time he retraced his steps he was mightily late to the New Year's Eve party to which he had been invited, and he made his apologies by standing in the doorway of the dining room and loudly asking Mrs. Dobzhansky to please pardon him his "behind," which brought down the by then well-lubricated house.

My father was a visitor as well as one visited, especially during his later years. There were trips to Brazil and other portions of Latin America, to Australia, India, and Egypt, as well as many meetings and lectures in Europe. He never succeeded in getting back to the

Soviet Union, although he tried several times. The nearest he came was the top of a hill in Finland, from which he could catch a glimpse of his native country.

Many summers were spent at a cabin in Mather, California, one of three botanical field stations run by Stanford University and Carnegie. The sea-level one was in Stanford; at an elevation of about 5,000 feet on the western boundary of Yosemite National Park was Mather, the intermediate; and on the eastern boundary of the park, at about 10,000 feet, was Timberline, on the crest of the Sierras. What started out as a convenient setup for the study of the effect of altitude on *Achillea* and *Potentilla* was equally useful for *Drosophila* experiments. The Stanford station was dropped, as being too far, and the bottom of the transect became Jacksonville, an abandoned gold-mining town at the foot of the mountains. There were several other collecting localities between Jacksonville and Mather, and Mather and Timberline, but Mather was the base for all the experiments and housed the laboratory.

The cabin at Mather consisted of one room with a front porch for sleeping and a screened-in back porch for cooking and eating. It was one of two cabins, but the lower one was always used by the botanists. There was no electricity, and there was cold running water only in the kitchen and the shower tent. A warm shower was possible if you built a fire under some coils of pipe, but that could only be done at certain hours of the day, given the constant danger of forest fires. The food for *Drosophila* and for humans was cooked on a wood-burning stove on the back porch. The cooking routine was arranged so that my mother cooked lunch and dinner, and my father and I and any assistants present took turns cooking breakfast and washing the day's dishes. French toast was my father's breakfast specialty. The bears, which were usually bears that had gotten into bad habits in the national park and had been trapped and then released near Mather, or so we all firmly believed, were the responsibility of my mother. When they periodically broke into the kitchen at night and wreaked havoc among the canned goods on the shelves and the food in the cooler cupboard, my father would resolutely refuse to wake up, and my mother had to frighten them off by herself.

Mather also gave my father a chance to indulge in his passion for horseback riding. For years he had a contract with a small stable down the road, which gave us horses five mornings a week. We rode mostly in Yosemite Park, the border of which was not far from the cabin. It was an obscure corner of the park, and it was a rare occasion when we met another human being. High points of several summers were trips taken to more distant and higher sections of the park, with four horses, one each for my father, myself, the cowboy who took care of the animals, and the pack animal to carry the week's supplies.

Life in Mather was not all fly-collecting and horseback riding. It was the site of several informal genetic conferences, during which the participants slept in sleeping bags under the pine trees and held their discussions under the same pine trees during the day. E. B. Ford attended one of these meetings and charmed everyone with his question, when first confronted with a sleeping bag, as to whether one slept on it, or in it and his subsequent triumphant adaptation to this new environment.

Ledyard Stebbins was a frequent visitor to Mather and used to go on horseback rides with us. I remember some passionate discussions about the hybrid and introgressive status of the manzanita bushes our horses were passing, starting with the gray-leaved *Arctostaphylos manzanita* near the cabin and gradually changing to the shiny green foliage of *Arctostaphylos patula* as the trail climbed past the gigantic sugar pine into the canyon that led to the park gate.

The gigantic sugar pine was struck and killed by lightning some time after I got married and stopped visiting Mather, but my parents continued to go there, and the shelves of the old cabin continued to be loaded with *Drosophila* bottles. Both my parents were fond enough of the place to ask that their ashes be buried there, even though their favorite grove of Ponderosa pines had been destroyed by a forest fire that had come uncomfortably near the cabin. My father deposited my mother's ashes after she died of a heart attack in February 1969, and I buried his ashes near the same granite boulder after his death in December 1975.

No account of my father's career could be complete without a mention of his collaborators. Boris Spassky started working for my father at Caltech, and he and his wife, Natalie, and their daughter made the trip east with us in 1940. Spassky was a Russian who had been trained as a forester and had escaped from the Soviet Union via Harbin in China. Until my father moved to Davis in 1971, Spassky and his wife were the pillars of stability in the laboratory, keeping the work going no matter what comings and goings were occurring around them. They were also dear friends, always invited to our house for holidays, especially the Russian ones, as we were invited to their house.

The other collaborator, also a Russian, was Olga Pavlovsky. My husband used to tease his parents-in-law, saying he was told by them that they didn't really know any Russians, whereas it was seldom that he heard English around them. Olga Pavlovsky moved to Davis with my father and kept the tradition of the Russian-speaking laboratory alive there.

After my mother died, the laboratory became for my father more and more what it had always been to a certain extent, his family. He would visit us occasionally, and take long walks at our farm, even though he considered the scenery of the eastern United States inferior. He was always interested in anthropology and would want to know the latest news, especially in physical anthropology and human evolution. But his own science was really his life, so that when Ernst Mayr wrote me recently saying that my father always considered himself a Russian, I demurred. Petty questions of nationality aside, the only country my father was ever a citizen of was a country that knows no boundaries, the country of science.

PART ONE
RUSSIAN ROOTS

■

Dobzhansky and Russian Entomology: The Origin of His Ideas on Species and Speciation

Nikolai L. Krementsov

ONE OF Dobzhansky's main achievements in the development of evolutionary theory, in the opinion of almost all historians, was his introduction of the biological species concept. We can see the essence of the concept in his own words (Dobzhansky 1935, p. 354):

> Considering dynamically, species represents that stage of evolutionary divergence, at which the once actually or potentially interbreeding array of forms becomes segregated into two or more separate arrays which are physiologically incapable of interbreeding.
>
> The fundamental importance of this stage is due to the fact that it is only the development of the isolating mechanisms that makes possible the coexistence in the same geographic area of different discrete groups of organisms.

This concept forms the core of Dobzhansky's classic 1937 book, *Genetics and the Origin of Species*, and of its final chapter, "Species as Natural Units." There, quoting himself, he argues that the origin of species is the stage "at which the once actually or potentially interbreeding array of forms becomes segregated into two or more separate arrays which are physiologically incapable of interbreeding" (p. 312).

In my paper, I will try to show how Dobzhansky's background in Russian entomology influenced this species concept.

Dobzhansky began his scientific career as an entomologist. From 1917 through 1935, he published more than thirty entomological articles. Western historians often overlook this fact, for obvious reasons: almost all of these papers were published in Russian, and Dobzhansky himself cited very few of them (and very few Russian entomological papers altogether) in his 1937 book. Thus, the role of Russian entomological theory and practice in the development of Dobzhansky's evolutionary thinking has been ignored. However, in my opinion the principal features of Dobzhansky's approach to evolution, and especially his species concept, are a direct continuation and development of ideas that were intensively discussed by Russian entomologists when Dobzhansky was among them.

THE RUSSIAN ENTOMOLOGICAL COMMUNITY

What did Russian entomology look like when the young Dobzhansky was a student? In the first quarter of the twentieth century, both institutionally and intellectually, entomology was one of the most developed biological disciplines in Russia.

The first Russian entomological society was organized in St. Petersburg in 1860; by 1917 it had almost five hundred members, living in many different cities and regions of the Russian empire. In the 1910s local entomological societies were organized in Lodz, Kiev, Moscow, Poltava, Stavropol, and other cities. The Russian Society of Applied Entomologists [*Rossiiskoe obshchestvo deiatelei po prikladnoi entomologii*] was organized in Kiev in 1913. The Moscow Entomological Society was created the next year. In addition, there were a number of special entomological laboratories, stations, and museums which were organized by various state and private authorities (e.g., the Ministry of Agriculture and State Properties, the All-Russian Society of Sugar Manufacturers, the Academy of Sciences, and various universities).

There were also many special entomological magazines and journals, published not only in Russian but often with titles, summaries, and sometimes entire articles in French, English, or German. Here I list only few of them, together with the year they began publishing and the city.

1861 (St. Petersburg) *Trudy Russkogo Entomologicheskogo Obshchestva* [Studies of the Russian Entomological Society]

1894 (St. Petersburg) *Trudy Biuro po Entomologii Uchenogo Komiteta Ministerstva Zemledelia i Gosudarstvennykh Imushchestv* [Studies of the Bureau of Entomology of the Scientific Committee of the Ministry of Agriculture and State Properties]

1901 (St. Petersburg) *Russkoe Entomologicheskoe Obozrenie* [Russian Entomological Review]

1912 (Kiev) *Entomologicheskii Vestnik* [Entomological Bulletin]

1914 (Kiev) *Vestnik Russkoi Prikladnoi Entomologii* [Bulletin of Russian Applied Entomology]

1915 (Moscow) *Izvestiia Moskovskogo Entomologicheskogo Obshchestva* [Proceedings of the Moscow Entomological Society]

Almost all of these journals continued to publish after the Bolshevik revolution (sometimes under new names), and both the number of such entomological journals and the size of their editions increased continually until 1928. For example, in 1923 the *Russian Entomological Review* was published in an edition of 200 copies; two years later, in 1925, the number had increased sixfold to 1200 copies. In addition, entomological papers were often published in other biological journals.

Perhaps the most important feature of the Russian entomological community at this time was the way it functioned *as a community*, forming a complex network of interconnections between professionals and amateurs, metropolitan and provincial investigators, elderly naturalists and young students. Of course, there were many special entomological meetings and conferences held during these decades, attended by most entomologists (such as the All-Russian Entomophytopathological Congress, which met almost every year beginning in 1918).[1] But underlying these formal gatherings was a rich tapestry of informal connections. Between and among entomologists throughout the empire, from its sophisticated capital through its smallest villages, there was a wide exchange of letters, articles, books, insect specimens, and so forth. It was a real scientific network, organized in such a way that students like Dobzhansky, who were beginning their scientific careers, were included in the intellectual atmosphere and institutional structure of the community.

We know, for instance, that in 1920 Dobzhansky participated in the zoological section of the Natural Sciences Department of the Ukrainian Scientific Society (which subsequently became the All-Ukrainian Academy of Sciences). On 12 September 1920 he gave a talk on "Some problems of the origin of genera of lady-beetles (Coccinelidae)."[2] In the early 1920s Dobzhansky corresponded with many well-known entomologists (G. G. Iakobson, N. N. Bogdanov-Kat'kov, V. N. Luchnik, and others). I found some books with Dobzhansky's signature and notes in the library of the Zoological Institute of the Academy of Sciences of the USSR. For example, an issue of the *Ukrainian Zoological Journal* that contains an article by Dobzhansky (1921a) is inscribed: "To G. G. Iakobson from the author."[3] The next issue of the

---

[1] See *Biulleteni Postoiannogo Biuro Vserossiiskikh Entomofitopatolicheskikh S"ezdov* [Bulletins of the Permanent Bureau of Entomophytopathology Congresses] (Petrograd, 1918–1923).

[2] See *Zoologicheskii Zhurnal Ukrainy*, 1921, no. 1, p. 28.

[3] Georgii G. Iakobson (1871–1926) was a prominent and influential Russian coleop-

same journal has another inscription: "To the Library of the Zoological Museum of the Russian Academy of Sciences from T. Dobrzhanskii." It should be added that Dobzhansky was elected a member of the Russian Society of Applied Entomologists in 1922, when he was only twenty-two years old.[4]

ENTOMOLOGICAL PRACTICE: THE EXCHANGE SYSTEM

Many of Dobzhansky's achievements in the development of evolutionary theory, it seems to me, are closely related to Russian entomological practice—the ways in which insects were collected, exchanged, and studied.

First of all, as a rule, entomological collecting yielded many specimens. Using the so-called mowing system of collection (in which an area is swept with a net for all insects present), vast numbers of insects of many different kinds were usually collected together, then sorted out by type. For example, on his expedition to Central Asia in the summer of 1925, Dobzhansky collected more than 10,000 specimens. Most insect species have great population density, so even primitive equipment (such as butterfly nets) can catch dozens and even hundreds of specimens of the same species from a small territory. Even if individuals of a species live relatively solitarily, entomologists may achieve the same result by using special traps that attract insects by light, food, water, etc. Numerous books instructing beginners how to catch insects were published in the 1910s and 1920s.[5] Articles describing new methods of collecting and new types of traps were regularly published in the entomological magazines.

As a rule, all entomologists had their own private collections of specimens that they had caught themselves. These included the types they were most interested in (e.g., beetles or butterflies). But the collecting process also produced vast numbers of other kinds of insects. This surplus provided the local entomologist with material for barter: a butterfly collector from Moscow, for example, could exchange beetles collected there for the butterflies a beetle specialist had collected elsewhere—say in Kiev or on the Volga.

---

terist. His major work (Iakobson 1905–1915) was a 1,000-page survey of the beetles of Eurasia On his life history, see Semenov-Tian'-Shanskii 1928.

[4] See "Spisok chlenov Rossiiskogo obshchestva deiatelei po prikladnoi entomologii" [The list of members of the Russian Society of Applied Entomologists], *Izvestiia Otdela Prikladnoi Entomologii*, 1922, vol. 2, p. 6.

[5] See, e.g , Korotnev 1914.

In the early decades of this century, the exchange of specimens between entomologists living in different regions was very popular. Advertisements for such exchanges were commonplace in contemporary Russian entomological magazines and journals. For instance, Viktor Luchnik, Dobzhansky's first adviser in entomology,[6] regularly published in the *Russian Entomological Review* the following notice: "V. N. Luchnik (Stavropol-Caucasian, City Museum) will send entomological material from North-Caucasian fauna in exchange for beetles of families Cicindelidae and Carabicidae."[7]

This practice of exchange produced private collections with many specimens of the same families or genera captured from many different regions. Often private collections fed into vast public ones through gifts and bequests to entomological societies and museums, most especially the Zoological Museum of the Russian Academy of Sciences. For instance, in 1914 A. P. Semenov-Tian'-Shanskii donated his own private collection, containing almost 100,000 specimens, to the Zoological Museum.[8] Every entomologist was able to study the Zoological Museum's vast treasure of insect specimens. Dobzhansky spent much time working with this collection when he lived in Leningrad, and almost all of his own collections ended up there. Most of the specimens Dobzhansky collected in the Kiev region (1918–1919) and on his various Russian expeditions (1925–1927) are still in this collection. Dobzhansky donated his own collection of Coccinelidae (about 13,000 specimens!) to the Zoological Institute.[9] Furthermore, even when Dobzhansky lived in the United States, he continued to send many specimens of different insects to the Zoological Museum. Dobzhansky's last gift to the collection was received on 14 November 1937.[10]

Unlike ornithologists or mammalogists, then, entomologists dealt

[6] Viktor N. Luchnik (1892–1936) was a member of the Russian Entomological Society from 1909. The first new species of ladybird beetle described by Dobzhansky was named in Luchnik's honor—*Coccinella lutchniki* (see Dobzhansky 1917). On Luchnik's life history see Plavil'shchikov 1938.

[7] E.g., in *Russkoe Entomologicheskoe Obozrenie*, 1910, vol. 10, no. 2, on the back inside cover.

[8] Andrei P. Semenov-Tian'-Shanskii (1866–1942) was president of the Russian Entomological Society from 1914 to 1933. On his life, see Iakobson 1920.

[9] See *Kniga postuplenii v Zoologicheskii Institut Akademii Nauk SSSR za 1935 god* [Book of acquisitions to the Zoological Institute of the USSR Academy of Sciences for 1935], p. 35.

[10] See *Kniga postuplenii v Zoologicheskii Institut Akademii Nauk SSSR za 1937 god* [Book of acquisitions to the Zoological Institute of the USSR Academy of Sciences for 1937], p. 59.

not with single specimens but with groups, series, masses of specimens. In such circumstances, the central practical and theoretical problem was to distinguish distinct species, and within them the subspecies or varieties. This was especially true when entomologists studied common, widespread species—such as Dobzhansky's early love, ladybird beetles.

I think it likely that Dobzhansky's early entomological work helped him adopt populational thinking. This was already evident in a 1924 paper and even more so in the expanded version of it that he presented at the Fourth International Congress of Entomology in Ithaca (Dobzhansky 1924c; 1928). Just as important, however, his deep involvement in Russian entomology immersed him in its literature and its problematics.

ENTOMOLOGY AND THE SPECIES PROBLEM

From 1900 to 1926 the entomological community was Russia's central locus of discussions over the various theories of species and speciation. This should be specially emphasized, since other disciplines (e.g., anatomy, physiology, animal breeding) were more interested in other problems—adaptations, phylogenesis, mutations. And it was the Russian entomologists who initiated the wide discussion on the species problem among the broader community of Russian zoologists. This was rather different from what happened in the West. This may be illustrated by comparing the discussions on "biological" (or "physiological") species among Russian and among Western zoologists.

At the end of the nineteenth century, almost simultaneously, two famous entomologists—T. D. A. Cockerell (of the United States) and N. A. Kholodkovskii (of Russia)[11]—described what they named the "physiological" (Cockerell 1897) or "biological" (Kholodkovskii 1900) species. These names were used to describe groups of organisms that were morphologically very similar but biologically clearly different. The differences between such groups may occur in food preferences, life cycle, places of breeding, etc. The idea of biological species attracted little attention in the West and was not much discussed, even among entomologists.[12] But as I have detailed else-

[11] Nikolai A Kholodkovskii (1858–1921) was a prominent Russian entomologist and evolutionist. On his life history, see Smirnov 1981.

[12] See, e.g. the discussion at the 143d regular meeting of the Entomological Society

where (Krementsov 1989), this idea was widely discussed in the Russian literature, not only by entomologists but also by ornithologists (V. L. Bianki), ichthyologists (L. S. Berg), and others.

Furthermore, in Western entomological literature the biological species was discussed only as a taxonomic unit (in connection with such purely taxonomic problems as the criteria for species, species definition, etc.), whereas in the Russian literature it was also discussed as an evolutionary one (in connection with such evolutionary problems associated with species concepts as the origin of species, factors affecting speciation, etc.). In the prerevolutionary decade evolutionary species concepts were regularly discussed at the sessions of the Moscow Entomological Society.[13]

Apparently, in terms of the differences between the interests of American versus Russian investigators, the situation in ornithology was just the reverse of that in entomology. As was noted by the naturalist and ichthyologist David Starr Jordan, "Of all branches of science, we may say, that the one most advanced in its developments, most nearly complete in its conclusion, is that of the systematic study of American birds. No other group of Naturalists has made such extensive studies of individual or of group variations as ornithologists, who have dealt with American birds" (D. Jordan 1905). But contemporary Russian ornithologists were much more interested in taxonomic and faunistic problems than in evolutionary ones (see, e.g., Bianki 1905, Sushkin 1916).

Among Russian zoologists, the entomologists were the first to discuss the problem of internal species structure. These discussions (of the problem of the species, its structure and component subunits, and factors affecting speciation) spread among entomologists the idea of the population as a group of specimens inhabiting a certain restricted range and differing from other groups of specimens in various features. In Russia it was the entomologists who began treating species as a complex of different conspecific populations.

(Washington), which was held 18 April 1899, in *Proceedings of the Entomological Society of Washington (1896–1901)*, 1901, vol. 4, pp. 386–389.

[13] At the seventh session, A. Musselius gave a special report entitled "Review of Species Theories"; the next session discussed A. Kazanskii's report, "Individual color variability of species of the genus Colias from observations in the Vladimir region." See "Izvlecheniia iz protokolov obshchikh sobranii Moskovskogo Entomologicheskogo Obshchestva za 1914–1915 god" [Extracts from records of general meetings of the Moscow Entomological Society 1914–1915], *Izvestiia Moskovskogo Entomologicheskogo Obshchestva*, 1915, vol. 1, pp. xviii–xxxiii.

The best example of such a treatment is the article by entomologist Andrei Semenov-Tian'-Shanskii (1910), "The taxonomic borders of the species and its subdivisions." This article stimulated a wide discussion involving not only most entomologists (e.g., Avinov 1912, Kholodkovskii 1912) but also important ichthyologists, ornithologists, and other zoologists (e.g., Berg 1910, Sushkin 1916, Alferaki 1910). As one zoologist aptly noted just before the revolution, "now investigators pay most attention to studying the units subordinate to species" (Bianki 1916, p. 290). And entomologists were ahead of other zoologists in this. In the 1920s Russian entomologists continued the tradition of the earlier decade, publishing many articles devoted to internal species structure (e.g., Alpatov 1924, Bartenev 1926).

Isolation and Speciation

In the works of Russian entomologists, a species was treated as a complex system of different elements joined by their morphological similarity and by their general geographical ranges, and separated from other species by the morphological hiatus and especially by their incapacity to breed with other species.

In the late nineteenth century, entomologists created an interesting idea that synthesized the concepts of the morphological hiatus and the incapacity to interbreed: the use of the morphology of the sexual apparatus as a definitive, diagnostic species characteristic. This concept was extremely popular aat the turn of the century in Germany (e.g., K. Jordan 1896), and Russian scholars much improved it (Petersen 1904, Iakhontov 1915). It is no accident that Dobzhansky's first article was titled "On the female sexual apparatus of ladybird beetles" (Dobzhansky 1921b, 1924a), and one of his next papers was "The sexual apparatus of ladybird beetles as a character of species and groups" (Dobzhansky 1924b, 1926).

In my opinion, the long tradition of treating the incapacity to interbreed (by morphological differences in the structure of the sexual apparatus) as a *characteristic of species* influenced the particular ways entomologists developed the idea of isolation as a *cause of speciation*. Russian entomologists related to these ideas differently than Western entomologists. First of all, it seems that Western entomologists were generally less interested in discussing these ideas than were their Russian colleagues. For example, T. D. A. Cockerell was the only entomologist who participated in the wide discussion in the pages of *Science* (1905–1906) about the role of isolation in the evolutionary process, and even he used more botanical than entomological material in the discussion (Cockerell 1906). Second, entomologists everywhere tended to see the morphological differences in the sexual apparatus of different groups within a species as a mechanical barrier that could isolate its progeny and thus create a new species; but there was a divergence of opinion between Western and Russian scholars about the origin of these differences.

It should be mentioned that there were (and are) two well-established traditions in discussing isolation as a cause of speciation. The first, based on Moritz Wagner's views about geographical isolation, treated the geographical splitting of a species as the beginning of speciation and divergence. The second, based on George Romanes's views about physiological selection, treated the physiological differentiation of individuals of a species inhabiting the same region as the beginning of speciation and divergence.

The differences between these two traditions can be roughly shown by the following sets of propositions:

Wagner tradition	*Romanes tradition*
1. The primary factor of speciation is geographical segregation of species.	1. The primary factor of speciation is physiological differentiation of species.
2. Adaptation to different geographical conditions (climate, soil, water, biological circumstances, etc.) is a main cause of speciation.	2. Adaptation to different local conditions (climate, soil, water, etc.) is accompanied by physiological isolation.
3. Sexual isolation is a side effect of adaptation and a result of speciation.	3. Sexual isolation is a direct effect of adaptation and a primary cause of speciation.
4. Geographical isolation is a necessary condition for the origin of sexual isolation.	4. Geographical isolation is not a necessary condition for the origin of sexual isolation.

The differences between these two points of view had special import for entomologists because the particular character of the differences in the structure of the sexual apparatus (which they regarded as a fundamental criterion of a species and the embodiment of its sexual isolation) seemed to have no adaptive meaning. The concept of geographical isolation was more popular among Western naturalists, whereas Russian naturalists tended to prefer physiological isolation. This divergence of opinion was clearly reflected in the polemical debate between K. Jordan and the influential Russian lepidopterist Wilhelm Petersen.[14]

Jordan argued that differences in the structure of the sexual apparatus between different species have no adaptive meaning and hence originated as a side effect of geographical isolation (K. Jordan 1903, 1905). By contrast, Petersen argued for a broader conception of *sexual apparatus* that would include not only the copulative organs but also related organs, for example those that help individuals of different sexes to recognize one another. He developed an example: Smell plays an important role in sex recognition among insects, but an insect's smell can depend on the chemical compounds in the plants on which it feeds. Hence, he proposed, if some members of an insect species were to change the plants on which they feed, this might result in a change in their smell; and this, in turn, might result in the "primary physiological isolation" of the individuals who eat one kind of plant from those who eat another, in the absence of any geographical isolation of the one kind from the other. Thus, the origin of physiological isolation may have an obvious adaptive character (Petersen 1903, 1904, 1909).

In light of this controversy, we should note that Dobzhansky's biological species concept was based not on geographical isolation but on the idea of *isolating mechanisms*. When discussing the problem of species in 1935, for example, he explicitly stated: "The emphasis should be placed however not on the absence of actual interbreeding between the different form complexes [i.e., geographical isolation—N. K.], but rather on the presence of physiological mechanisms making interbreeding difficult or impossible" (Dobzhansky 1935, p. 349). Three decades earlier, almost exactly the same words had been used by Petersen (Petersen 1903).

Petersen's ideas were developed and elaborated by A. Semenov-Tian'-Shanskii. "I have decided," Semenov wrote, "that physiological isolation is an important factor in speciation and, hence, isolation is an important criterion for species definition" (Semenov-Tian'-Shanskii 1910, p. 14). In this same article, drawing on Petersen's works, Semenov expanded the notion of isolation, giving a lengthy classification of types, notably: *seasonal,* or *chronological,* isolation; *psycho-physiological,* or *sexual,* isolation (today we would say behavioral, or ethological); and *mechanical* isolation.

[14] On Petersen's life history, see Kuznetsov 1937.

Semenov's ideas on isolation, and his classification of different kinds of isolation, exercised a pervasive influence on Russian investigators in many fields. Sergei Chetverikov drew upon them in his famous 1926 paper; so, too, did Alexander Serebrovskii in his book *The Hybridization of Animals* (1935). This background helps us understand the historical sources of a key part of Dobzhansky's 1937 classic, *Genetics and the Origin of Species*. Its eighth chapter, "Isolating Mechanisms" (pp. 228–258), is subdivided into sections distinguishing "ecological and seasonal isolation," "sexual isolation," and "mechanical isolation." Furthermore, these sections are introduced by a chart detailing the various kinds of "physiological isolating mechanisms" (pp. 231–232) which shows the strong influence of Semenov-Tian'-Shanskii's ideas.

I would like to emphasize again that in their discussion of the problem of speciation, Russian entomologists treated physiological isolation as a central factor. As another famous Russian entomologist, Nikolai Kuznetsov,[15] wrote in 1917, "The cause of the origin of biological groups [species] at first is a sexual splitting of groups, which marks the beginning of them [species], therefore the notion of species coincides with the notion of sexual isolation. Consequently, the essence of the species idea is the idea of sexual isolation" (Kuznetsov 1917, p. 79). No wonder that a decade later Chetverikov wrote: *"The real source of speciation, the real cause of the origin of species is not selection, but isolation"* (Chetverikov 1926, p. 42).

We should note that Dobzhansky's subsequent development of the concept of isolation sought to reconcile the controversy between the ideas of geographical and physiological isolation. By proposing that isolating mechanisms may originate by selection against the hybrids of different geographical arrays of a species, Dobzhansky was able to show that physiological isolation itself may have adaptive significance.

In my opinion, then, the essence of the biological species concept was clearly formulated in the theoretical papers of leading Russian entomologists in the prerevolutionary decades. This was the place from which both Chetverikov and Dobzhansky mined the cornerstones of their ideas of speciation.

[15] Nikolai Ia. Kuznetsov (1873–1948) was a prominent Russian lepidopterist and insect physiologist. He was vice-president of the Russian Entomological Society from 1933 to 1948. On his life history, see *Entomologicheskoe Obozrenie*, 1949, vol. 30, no. 3–4; the entire issue was dedicated to his memory.

Entomological Roots of Russian Genetics

It is especially significant that the growing community of Russia's geneticists had close connections with the entomological community. In the 1910s almost all Russians who were doing genetics belonged to various entomological circles and local societies.

In Moscow, for example, Sergei Chetverikov—the founder of Russian population genetics—was a member of St. Petersburg's Entomological Society from 1910 and a founder of the Moscow Entomological Society in 1914 and chief editor of its bulletin. The Moscow Entomological Society included in its membership the city's leading geneticists: Nikolai Kol'tsov, Maria Sadovnikova-Kol'tsova, Anna Chetverikova, and Alexandr Serebrovskii.[16] Dobzhansky's teacher in entomology, Viktor Luchnik, was also a member. Among other members of the society from Dobzhansky's generation were Dmitrii Romashov and Vladimir Alpatov, two geneticists who did a lot of entomology in the 1920s.

St. Petersburg's geneticists also had close relations with entomology. Iurii Filipchenko—Dobzhansky's mentor in genetics—was a member of the Russian Entomological Society from 1907, published entomological papers, and did collaborative work with the influential Petersburg entomologist Nikolai Kholodkovskii (Filipchenko 1916, 1926). And after Dobzhansky arrived in the city to work at Filipchenko's genetics laboratory, he continued his early entomological studies in cooperation with none other than the president of the Entomological Society—and the "classifier" of physiological isolating mechanisms—A. P. Semenov-Tian'-Shanskii (Semenov-Tian'-Shanskii and Dobrzhanskii 1923, 1927).[17] Dobzhansky may have "infected" Filipchenko's other pupils with his entomological interest. For instance, Iulii Kerkis participated in Dobzhansky's entomological work in Central Asia and published an article on the sexual apparatus of Hemiptera in which he thanked Dobzhansky for advice (Kerkis 1926).

The close personal connections between Russia's animal geneticists and its entomological community during the 1910s and 1920s

---

[16] "Sostav Moskovskogo Entomologicheskogo obshchestva k 1 maia 1915 goda" [A list of members of the Moscow Entomological Society as of 1 May 1915], *Izvestiia Moskovskogo Entomologicheskogo Obshchestva*, 1915, vol. 1, pp. viii–xi.

[17] One of the new species described by Dobzhansky was probably named in honor of A. P. Semenov-Tian'-Shanskii—*Coccinella tianshanica* (Dobzhansky 1927).

stimulated a mutual interest by each group in the problems of the other. In this light, it is significant that Dobzhansky's first genetics paper was devoted to the changes in the sexual apparatus of some mutant *Drosophila melanogaster* (Dobzhansky 1924d).

Unlike the Russian geneticists, the American geneticists did not have such close connections with entomologists. Nobody from T. H. Morgan's genetics school had the kind of entomological experience that Dobzhansky did. And I suppose it is no accident that the first American article published by Dobzhansky (in cooperation with C. B. Bridges) was entitled "The reproductive system of triploid intersexes in *Drosophila melanogaster*" (Dobzhansky and Bridges 1928). Perhaps Dobzhansky "infected" not only some Russian but also some American geneticists with his entomological approach to the species problem.

Among Russian zoologists, it was the entomologists who best understood the necessity of a genetic interpretation of the species problem. In 1903 Petersen proposed that differences in the structure of the sexual apparatus may have originated from some variations in the "germ plasm" (Petersen 1903). In 1912 Kholodkovskii suggested that the origin of the two biological species of *Chermes* that he had discovered may have been based on their different number of chromosomes, which made them incapable of interbreeding (Kholodkovskii 1912). Filipchenko, who studied these biological species statistically, supported this idea (Filipchenko 1916). Subsequently, the founder of Moscow genetics, Nikolai Kol'tsov, discussed this as a mode of speciation in a theoretical review (Kol'tsov 1922).

In 1917 the entomologist Nikolai Kuznetsov published a portentous observation. Geneticists, he noted, had been studying the same intraspecific units that had been delineated by insect systematists, although geneticists had not used the taxonomic terminology. And then he wrote: "It would be good if descriptive taxonomists carefully learned the theories of heredity and evolution, and tried to coordinate taxonomic facts with the results of experimental genetics studies and, accordingly, to change the taxonomic terminology" (Kuznetsov 1917, p. 76).

In a sense, Kuznetsov's words outlined a scientific research program—one that would be begun by the founders of Russian genetics (Kol'tsov, Chetverikov, and Serebrovskii) but that would be most fully realized in the researches that Dobzhansky pursued throughout his life.

ACKNOWLEDGMENTS

I am grateful to many institutions and individuals who have helped me with this study. First, thanks go to my colleagues in the Institute of the History of Science and Technology for helpful discussions. I am grateful to the staff of the Zoological Institute (St. Petersburg), and especially to I. M. Kerzhner, for help in locating and studying Dobzhansky's entomological collections. Finally, I wish to express my profound gratitude to Mark Adams and Daniel Todes for their friendly criticisms and for their patience in helping me with English.

REFERENCES

Alferaki, S. 1910. "Neskol'ko soobrazhenii po povodu truda A. P. Semenova-Tian'-Shanskogo—'Taksonomicheskie granitsy vida i ego podrazdelenii'" [Some comments on A. Semenov-Tian'-Shanskii's paper—"The taxonomic borders of the species and its subdivisions"]. *Biologicheskii Zhurnal*, vol. 1, pp. 165–170.

Alpatov, V. V. 1924. "Izmenchivost' i nizshie taksonomicheskie kategorii. K sistematike murav'ev" [Variability and lower systematic categories. Toward the systematics of ants]. *Russkii Zoologicheskii Zhurnal*, vol. 5, no. 1–2, pp. 227–244.

Avinov, A. 1912. "O nekotorykh formakh roda *Parnassius*" [On several forms of the genus *Parnassius*]. *Trudy Russkogo Entomologicheskogo Obshchestva*, vol. 40, no. 5.

Bartenev, A. N. 1926. "O nizshikh taksonomicheskikh edinitsakh" [On lower taxonomic units]. *Izvestiia Severo-Kavkazskogo Universiteta*, vol. 11, pp. 8–38.

Berg, L. S. 1910. "O vide i ego podrazdeleniiakh" [On the species and its subdivisions]. *Biologicheskii Zhurnal*, vol. 1, pp. 1–7.

Bianki, V. L. 1905. "Dopolnitel'nye zametki o palearkticheskikh zhavoronkakh (*Alaudidae*)" [Supplementary notes on paleoarctic skylarks]. *Izvestiia Imperatorskoi Akademii Nauk*, vol. 23, no. 3, pp. 205–240.

———. 1916. "Vid i podchinennye emu taksonomicheskie formy" [The species and its taxonomic subforms]. *Russkii Zoologicheskii Zhurnal*, vol. 1, pp. 287–297.

Chetverikov, S. S. 1926. "O nekotorykh momentakh evolutsionnogo protsessa s tochki zreniia sovremennoi genetiki" [On some aspects of the evolutionary process from the viewpoint of modern genetics]. *Zhurnal Eksperimental'noi Biologii*, ser. A, vol. 2, no. 1, pp. 3–54.

———. 1983. *Problemy obshchei biologii i genetiki* [Problems of general biology and genetics]. Novosibirsk: Nauka.

Cholodkovsky. See Kholodkovskii.
Cockerell, T. D. A. 1897. "Physiological species." *Entomological News*, vol. 8, pp. 234–236.
———. 1906. "The evolution of species through climatic conditions." *Science*, vol. 23, pp. 145–146.
Dobzhansky, Th. [Dobrzhanskii, F.] 1917. "Opisanie novogo vida roda *Coccinella* iz okrestnostei Kieva" [Description of a new species of the genus *Coccinella* from the neighborhood of Kiev]. *Materialy k poznaniiu fauny Iugo-Zapadnoi Rossii*, vol. 2, pp. 46–47.
———. [Dobrzhanskii, F.] 1921a. "On the fauna of the Coccinellidae of Volhynia and Podolia." *Zoologicheskii Zhurnal Ukrainy*, vol. 1, no. 1, pp. 20–23 (in Ukrainian).
———. [Dobrzhanskii, F.] 1921b. "K poznaniiu zhenskogo polovogo apparata bozh'ikh korovok" [On the female sexual apparatus of ladybird beetles]. *Kievskie Uchenye Zapiski*, vol. 1, pp. 1–3.
———. [Dobrzhansky, T.] 1924a. "Beitrag zur Kenntis des weiblichen Geschlechtapparates der Coccinelliden." *Zeitschrift für wissenschaftliche Insektenbiologie*, vol. 19, pp. 98–100.
———. [Dobrzhansky, T.] 1924b. "Die weiblichen Generationsorgane der Coccinelliden als Artmerkmal betrachtet." *Entomologische Mitteilungen*, vol. 13, no. 1, pp. 18–27.
———. [Dobrzhanskii, F.] 1924c. "O geograficheskoi i individual'noi izmenchivosti *Adalia bipunctata* L. i *Adalia decempunctata* L." [On geographical and individual variability of *Adalia bipunctata* L. and *Adalia decempunctata* L.]. *Russkoe Entomologicheskoe Obozrenie*, vol. 18, no. 4, pp. 201–212.
———. [Dobrzhanskii, F.] 1924d. "Über den Bau des Geschlechtsapparats einiger Mutanten von *Drosophila melanogaster* Meig." *Zeitschrift für Induktive Abstammungs- und Vererbungslehre*, vol. 34, pp. 245–248.
———. [Dobrzhansky, T.] 1926. "Polovoi apparat bozh'ikh korovok kak vidovoi i gruppovoi priznak" [The sexual apparatus of ladybird beetles as a character of species and groups]. *Izvestiia Akademii Nauk SSSR*, no. 13–14, pp. 1385–1394; no. 15–16, pp. 1555–1586.
———. [Dobzhanskii, F.] 1927. "Materialy dlia fauny Coccinellidae Semirech'ia" [Materials for the fauna of Coccinellidae of Semirech'e]. *Russkoe Entomologicheskoe Obozrenie*, vol. 21, pp. 43–52.
———. 1928. "The origin of geographical varieties in Coccinellidae." In *Transactions of the Fourth International Congress of Entomology—Ithaca, New York*, vol. 2, p. 563.
———. 1935. "A critique of the species concept in biology." *Philosophy of Science*, vol. 2, no. 3, pp. 344–355.
———. 1937. *Genetics and the Origin of Species*. New York: Columbia University Press.

Dobrzhansky, T., and C. B. Bridges. 1928. "The reproductive system of triploid intersexes in *Drosophila melanogaster*." *American Naturalist*, vol. 62, pp. 425–434.

Filipchenko, Iu. A. 1916 "Biologicheskie vidy khermesov i ikh statisticheskoe razlichenie" [Biological species of *Chermes* and their statistical differentiation]. *Zoologicheskii Vestnik*, vol. 1, no. 2, pp. 261–285.

———. 1926. "*Collembola*, sobrannye ekspeditsiei V. A. Dogelia i I. I. Sokolova v Britanskoi Vostochnoi Afrike" [On the *Collembola* collected by the expedition of V. A. Dogiel and I. I. Sokolov in British East-Africa]. *Russkoe Entomologicheskoe Obozrenie*, vol. 20, pp. 180–196.

Iakhontov, A. 1915. "Rassovye razlichia v stroenii muzhskogo polovogo apparata u nekotorykh *Lepidoptera—Rhopalocera*" [Racial differences in the structure of the male sexual apparatus of certain *Lepidoptera—Rhopalocera*]. *Izvestiia Moskovskogo Entomologicheskogo Obshchestva*, vol. 1, pp. 40–57.

Iakobson. G. G. 1905–1915. *Zhuki Rossii i Zapadnoi Evropy* [The beetles of Russia and Western Europe]. St. Petersburg: A. F. Devrien. 1024 pp.

———. 1920. "Entomologicheskoe otdelenie Zoologicheskogo Muzeia Rossiiskoi Akademii" [The entomological department of the Zoological Museum of the Russian Academy of Sciences]. *Biulleten' 2-go Vserossiiskogo entomofitopatologicheskogo s"ezda v Petrograde 25–30 oktiabria 1920 g.*, no. 6, pp. 19–21.

Jordan, D. S. 1905. "The origin of species through isolation." *Science*, vol. 22, pp. 545–562.

Jordan. K. 1896. "On mechanical selection and other problems." *Novitates Zoologica*, vol. 3, pp. 426–525.

———. 1903. "Bemerkungen zu Herrn Dr. Petersen's Aufsatz: Enstehung der Arten durch physiologische Isolierung." *Biologisches Zentralblatt*, vol. 23, pp. 660–664.

———. 1905. "Der Gegensatz zwischen geographischer und nichtgeographischer Variations." *Zeitschrift für wissenschaftliche Zoologie*, vol. 83, pp. 151–210.

Kerkis. Iu. 1926. "K poznaniiu vnutrennego polovogo apparata vodnykh *Hemiptera-Heteroptera*" [Toward an understanding of the internal sexual apparatus of water *Hemiptera-Heteroptera*]. *Russkoe Entomologicheskoe Obozrenie*, vol. 20. no. 3–4, pp. 296–307.

Kholodkovskii (Cholodkovsky). N. A. 1900. "Über den Lebenszyklus der Chermes-Arten und die damit verbundenen algemeinen Fragen." *Biologische Zentralblatt*, vol. 20, pp. 265–283.

———. 1912. "O biologicheskikh vidakh" [On biological species]. *Izvestiia St-Peterburgskoi Biologicheskoi Laboratorii*, vol. 12. no. 2–3, pp. 105–107.

Kol'tsov. N. K. 1922. "Obrazovanie novykh vidov i chislo khromosom" [The formation of new species and the number of chromosomes]. *Uspekhi Eksperimental'noi Biologii*, vol. 1, no. 2, pp. 181–196.

Korotnev, N. I. 1914. *Zhuki. Prakticheskie rukovodstvo k nauchnomu sobiraniiu i vospitaniiu zhukov i sostavleniiu kollektsii* [Beetles. Practical guide to scientific collecting and growing of beetles and to forming collections]. Moscow: I. D. Sytin. 176 pp.

Krementsov, N. L. 1989. "Evoliutsionnye aspekty povedeniia zhivotnykh (istoriko-kriticheskii analiz otechestvennykh issledovanii)" [Evolutionary aspects of animal behavior (a critical historical analysis of Russian investigations)]. Candidate dissertation. Institute of the History of Science and Technology, USSR Academy of Sciences. Moscow. 201 pp.

Kuznetsov, N. Ia. 1917. "O 'taksonomicheskikh poniatiiakh' i popytkakh ikh obosnovaniia morfologicheskimi dannymi" [On "taxonomic concepts" and on attempts to base them on morphological data]. *Russkoe Entomologicheskoe Obozrenie*, vol. 17, pp. 53–80.

————. 1937. "Pamiati Vil'gel'ma Erastovicha Petersena" [In memory of Wil'gel'm Erastovich Petersen]. *Entomologicheskoe Obozrenie*, vol. 27, no. 1–2, pp. 139–142.

Petersen, W. 1903. "Enstehung der Arten durch physiologische Isolierung." *Biologische Zentralblatt*, vol. 23, pp. 468–477.

————. 1904. "Die Morphologie der Generationsorgane der Schmetterlinge und ihre bedeutung für die Artbildung." *Zapiski Imperatorskoi Akademii Nauk*, vol. 16, no. 8, pp. 1–84.

————. 1909. "Ein Beitrag zur Kenntis der Cattung Eupithecia Curt. vergleichende Untersuchung der Generationsorgane." *Deutsche Entomologische Zeitschrift*, vol. 4, pp. 203–313.

Plavil'shchikov, N. 1938. "Luchnik V. N." *Entomologicheskoe Obozrenie*, vol. 277, no. 3–4, pp. 267–280.

Semenov-Tian'-Shanskii, A. P. 1910. "Taksonomicheskie granitsy vida i ego podrazdeleniia" [The taxonomic borders of the species and its subdivisions]. *Zapiski Imperatorskoi Akademii Nauk, fiziko-matematicheskoe otdelenie*, vol. 25, no. 1, pp. 1–29.

————. 1928. "Pamiati G. G. Iakobsona" [In memory of G. G. Iakobson]. *Russkoe Entomologicheskoe Obozrenie*, vol. 22, no. 1–2.

Semenov-Tian'-Shanskii, A. P., and F. Dobrzhanskii. 1923. "Tri novykh vida semeistva *Coccinelidae (Coleoptera)* iz Aziatskoi Rossii" [Three new species of the family *Coccinelidae (Coleoptera)* from Asian Russia]. *Russkoe Entomologicheskoe Obozrenie*, vol. 18, no. 2–3, pp. 99–102.

————. 1927. "Lichinka *Silphopsyllus desmanae* Ols., zhuka-parasita vykhukholi, kak kriterii ego geneticheskikh otnoshenii i sistematicheskogo polozheniia" [The larva of *Silphopsyllus desmanae* Ols., the parasitic beetle of the muskrat, as a criterion of its genetic connections and systematic position]. *Russkoe Entomologicheskoe Obozrenie*, vol. 21, pp. 8–16.

Serebrovskii, A. 1935. *Gibridizatsia zhivotnykh* [The hybridization of animals]. Moscow and Leningrad: Biomedgiz.

Smirnov, O. V. 1981. *N. A. Kholodkovskii.* Moscow: Nauka.

Sushkin, P. 1916. "Podvid (*subspecies*) i plemia (*natio*)" [Subspecies and tribe]. *Ornitologicheskii Vestnik*, vol. 7, no. 4, pp. 204–208.

■

Filipchenko and Dobzhansky:
Issues in Evolutionary Genetics
in the 1920s

Daniel A. Alexandrov

DOBZHANSKY always regarded Filipchenko as his mentor, but it is far from obvious what Dobzhansky's population genetics and Filipchenko's quantitative studies of variation and inheritance had in common. To modern geneticists, Filipchenko's researches look remarkably old-fashioned, even by the standards of his own day. If we look more closely at the works of these two men during the first quarter-century, however, we can uncover common ground. Analyzing their similarities and differences not only helps us understand the nature of Dobzhansky's debt to Filipchenko but also clarifies the nature of Dobzhansky's later work.

IURII FILIPCHENKO

Iurii Filipchenko (1882–1930) belongs to the first generation of Russian geneticists.[1] Educated as a classical zoologist, he graduated from the Zoology Department of St. Petersburg University in 1905. His first research project, which led to his candidate thesis (1913), dealt with the comparative embryology of arthropods. As a comparative embryologist, Filipchenko was interested in the problem of the manifestation and evolutionary development of characteristics, and his comparison of the developmental patterns in higher taxa (orders, subclasses, and classes) gave him a broad perspective. This was the level of evolution that Filipchenko later termed *macroevolution*.

Although Filipchenko's research was always narrowly focused on various particular organisms and problems, we should remember

---

[1] The most complete and informative biography of Filipchenko was written by Mark B. Adams, "Filipchenko [Philiptschenko], Iurii Aleksandrovich," in *Dictionary of Scientific Biography*, vol. 17, Supp. 2, ed. Frederic L. Holmes (New York: Charles Scribner's Sons, 1990), pp. 297–303. The most complete biography in Russian is N. N. Medvedev, *Iurii Aleksandrovich Filipchenko* (Moscow: Nauka, 1978).

that he had an encyclopedic and commanding knowledge of much of biology. He was the first in Russia to give a special university course on the problems of inheritance and also the first to found a separate university department in that field. His textbooks on inheritance, methods for studying variation, and genetics were very influential. He was undoubtedly the most prolific Russian writer of his day in this field. Dobzhansky and many other Russian biologists first became acquainted with modern genetics through Filipchenko's articles and books. With good reason, the Soviet botanist and historian Zhukovskii has called Filipchenko "the teacher of our youth."

Like many other Petersburg biologists of his generation, Filipchenko was deeply influenced by vitalism and holism. His philosophical approach can be characterized as a kind of organicism. In his candidate thesis, he noted: "The process of the evolution of organisms can be explained neither by so-called Lamarckian factors, nor by selection—it is one of the basic features of living beings. Even allowing the possibility of reducing a series of life processes to purely mechanistic causes, this will not allow us to explain the life of an organism as a whole."[2]

In his early investigations on inheritance, Filipchenko focused on quantitative variations, regarding this as the most effective way to investigate the inheritance of natural characters of the organism as a whole. In his doctoral dissertation on the inheritance of skull shape in mammals, he cited works by Johannsen, Castle, and other geneticists who interpreted quantitative inheritance in Mendelian terms. In the 1910s, then, Filipchenko was a Mendelist, but a Mendelist who took the best from biometrics: for him there never was any contradiction between Mendelian and biometric approaches to heredity. He always regarded both Mendel and Galton as the founders of genetics.

Variation

Filipchenko and his students studied, described, and compared quantitative variation within and among many groups of organisms,

---

[2] See D. A. Aleksandrov, "Iurii Aleksandrovich Filipchenko kak genetik-evoliutsionist: formirovanie nauchnykh interesov i vzgliadov" [Iurii Aleksandrovich Filipchenko as an evolutionary geneticist: the formation of his scientific interests and views], in *Evoliutsionnaia genetika* [Evolutionary genetics], ed. S. G. Inge-Vechtomov (Leningrad: Leningrad University Press, 1982), pp. 3–21.

sexes, species, small and large genera, and so forth.[3] Because of Filipchenko's deep distaste for any form of ectogenesis and especially Lamarckism, however, these studies of variation had no connection with the problem of adaptation.

A central feature of Filipchenko's approach to variation was his distinction between *individual variation* and *group variation*. He defined individual variation as noninherited variability caused by the influence of the environment; he used this term roughly as a synonym for *modifications*. By contrast, group variation he regarded as variation of inherited characteristics; he defined it as variation in the "composition of each species of Jordanons and biotypes, considering biotypes as groups of individuals with identical genotypes (following W. Johannsen)."[4] For Filipchenko, then, mutations were changes of old biotypes and the appearances of new ones.

I want to emphasize that this image of the hierarchical organization of variation is quite different from genetic variation within free-mating populations based on a shuffling of genes. Of course, Filipchenko knew all about work on the shuffling of genes and gene combinations in free-mating populations and discussed it in detail in the textbook cited above in its final chapter, entitled "the mathematical basis of breeding theory." Nonetheless, his thinking on variation was dominated by his hierarchical view on the organization of species and taxa.

Inheritance

Filipchenko understood two kinds of inheritance: Mendelian inheritance of variation within species and non-Mendelian (and nonchromosomal) inheritance of variation in macroevolutionary characters delineating higher taxa. He spent his most productive years in studies of such hypothetical non-Mendelian inheritance in wheats. Regarding these studies, Dobzhansky would later remark: "He bet on the wrong horse."[5] But these wheat studies were quite revealing, for

[3] Filipchenko's treatment of variation and its relation to evolution is treated in most of his books 1917–1929, most notably *Izmenchivost' i metody ee izucheniia* [Variation and methods for its study] (Moscow and Leningrad: Gosizdat), which went through four editions, 1923–1929. See also Ju. Philiptschenko, *Variabilität und Variation* (Berlin: Gebrüder Borntraeger, 1927).

[4] Iu. Filipchenko, *Genetika* [Genetics] (Moscow and Leningrad: Gosizdat, 1929), p. 93. He gave the same definition in his other works.

[5] Theodosius Dobzhansky, 1962–1963. "The Reminiscences of Theodosius Dobzhan-

their central concern was the effects of genes on development—what we would now call phenogenetics. Filipchenko's reason for studying wheats in the late 1920s are sometimes attributed simply to their practical, agricultural importance. But Filipchenko gave another reason for studying such inheritance in plants: they afforded the opportunity for studying the effects of gene activity unobscured by the complicating effects of the endocrine system.[6] He regarded studies of the effects of gene action on development to be a new means of clarifying the relationship between inheritance and evolution.

Evolution

Filipchenko's interest in nongenic inheritance was connected to his interest in the problem of macroevolution. In his 1927 German monograph *Variabilität und Variation*, he distinguished two independent levels of evolution: evolution within species, and the formation of Jordanons and Linneons, he termed *microevolution*; evolution above the species level, the formation of genera and higher taxa, he termed *macroevolution*. Microevolution is governed by the natural selection of characteristics whose inheritance can be understood in terms of genetics. Macroevolution, from Filipchenko's point of view, is purely internally driven and based on variability of cytoplasmic nongenic inheritance.

As is the case for many other zoologists, Filipchenko understood evolution in terms of historical changes in the morphological characteristics of organisms. A vision of the organism, heredity, and evolution in terms of morphological or morphophysiological characteristics dominated his thinking. He did regard mutations as a main factor in evolution, but he always thought of mutations in terms of "forms" rather than "genes" or "alleles." For example, in the third chapter of his 1929 textbook on genetics, he defined mutations as "sudden changes of genotype," but his discussion makes clear that he understood mutations as sudden changes of characteristics; he often used the phrases "sudden appearance of new forms, i.e. muta-

sky." Typed transcript. 2 parts. Oral History Research Office, Columbia University, New York.

  [6] For an example of his phenogenetic approach to plant genetics, see Jur. Philiptschenko, "Gene und Entwicklung der Ährenform beim Weizen," *Biologische Zentralblatt* vol. 49, no. 4 (1929), pp. 1–16.

tions" or "old and new forms."[7] Although a founder of Russian genetics, Filipchenko remained a zoologist in his attitude toward evolution and variation.[8]

Species and Their Hybrids

It may seem that Filipchenko would have regarded speciation as the watershed between micro- and macroevolution, but this is not the case. Filipchenko never considered speciation as the central problem of evolution. In his view, speciation was the same gradual process of differentiation of groups as any other microevolutionary change, governed by natural selection, "at first biotypes, then Jordanons and, finally, new species or Linneons."[9] For him, the distinction between Linnean species and Jordanons was only quantitative.

He regarded some cases of speciation as involving macroevolutionary changes in cytoplasmic, nongenic inheritance, of course, as when various species within the same genus differed in such inheritance; but he generally treated such changes as a result of their evolutionary divergence after speciation. This point of view is reflected in Filipchenko's discussion of interspecific hybridization and sterility.

From his earliest genetic papers, Filipchenko paid much attention to the problem of species hybrids. For him, the study of hybrids between species, or, even better, species of different genera, was an especially good way to analyze the nongenic inheritance of macroevolutionary characteristics. In the 1910s he intensively studied variation and inheritance in interspecific hybrids of mammals, mainly cattle and rodents,[10] and his first review in genetics was concerned with this problem.[11]

[7] Filipchenko, *Genetika*, p. 106.

[8] While remaining a "zoologist" in his morphological view of variation and evolution, Filipchenko was nonetheless an "experimentalist" in his theoretical and epistemological attitudes. See the important discussion of this, comparing Filipchenko's and Severtsov's views, in Mark B. Adams, "Severtsov and Schmalhausen: Russian Morphology and the Evolutionary Synthesis," in E. Mayr and W. Provine, eds., *The Evolutionary Synthesis: Perspectives on the Unification of Biology* (Cambridge: Harvard University Press, 1980), pp. 193–225 and especially pp. 197–204.

[9] Iu. Filipchenko, *Evoliutsionnaia ideia v biologii* [The evolutionary idea in biology], 3d ed. (Moscow: Nauka, 1977), p. 190.

[10] Adams, "Filipchenko [Philiptschenko], Iurii Aleksandrovich," p. 299 (see Note 1).

[11] Iu. Filipchenko, "O vidovykh gibridakh" [On species hybrids], in *Novye idei v*

Throughout his career, Filipchenko stated his understanding of interspecific hybrid sterility as a by-product of divergence. He repeatedly emphasized that sterility gradually appears as species diverge: "Formerly it was considered that hybrids between species are sterile, while mongrels between races are fertile. The erroneousness of this is obvious from the fact that degrees of sterility in species hybrids are very diverse: if species are very close to one another their hybrids are highly fertile; if initial forms are more distant from one another, one sex in their progeny tends to be sterile (more often the male than the female); in hybrids between very distant species, both sexes are completely sterile."[12] In his early paper on "biological species" of insects (initially described by Russian entomologist N. Kholodkovskii on the basis of life cycle differences only), Filipchenko's whole point was to show that even such "biological" or "physiological" species or races can be distinguished as morphological forms with the help of certain biometrical techniques.[13] Thus, for Filipchenko hybrid sterility did not help to distinguish species from one another; only morphological or morphophysiological criteria were useful for this purpose, in Filipchenko's view, because a species is a "form."

Filipchenko's zoological view of variations, inheritance, evolution, and species provides the context for understanding Dobzhansky's views on these subjects.

THEODOSIUS DOBZHANSKY

Theodosius Dobzhansky, then Feodosii Grigor'evich Dobrzhanskii, began his scientific studies in entomology in his high school years under the influence of the Kiev entomologist Viktor Luchnik.[14] From his teenage years, Dobzhansky studied beetles. As we know, entomology can be an absorbing passion, and collecting and classifying beetles can be an especially obsessive and sophisticated business, exciting for beginners.[15]

---

biologii 4. *Nasledstvennost'* [New ideas in biology, 4. Inheritance], ed. V. Vagner (St. Petersburg: Obrazovanie, 1913), pp. 124–149.

[12] Filipchenko, *Genetika*, p. 311.

[13] Iu. Filipchenko, "Biologicheskie vidy khermesov i ikh statisticheskoe razlichenie" [Biological species of *Chermes* and their statistical differentiation], *Zoologicheskii Vestnik* vol. 1, no. 2 (1916), pp. 261–277; with French translation, pp. 277–285.

[14] In addition to published and other archival sources, my principal source on Dobzhansky's life is his oral memoirs. See Note 5.

[15] We are reminded of Charles Darwin's similar passion and of his remark "It seems

But Dobzhansky himself considered his first real teacher in zoology and biology to be the Kiev zoologist Sergei Kushakevich, a largely unknown figure who deserves more attention. Kushakevich graduated from Kiev University and worked for a time in Germany, where he developed a consuming interest in the new fields of embryology and cytology, the subjects of his candidate and doctoral dissertations at St. Petersburg University. His analysis of the development of the colonial algae *Volvox*, published posthumously, received a great response among Petrograd biologists.[16] His interests also extended to genetics and evolution. In 1914 he published an article in the popular science journal *Priroda* on chromosomes and inheritance and taught courses on "the doctrine of evolution" and "the doctrine of inheritance."[17] Entomological and cytological studies under Kushakevich's direction seem to have played a formative role in shaping Dobzhansky's interests in evolution and experimental biology.[18]

Dobzhansky's studies of variation in ladybird beetles are well known, in part because he continued these studies in the United States and often used them as examples in his books. His first work in this series was planned in Kiev and conducted in Leningrad during his first months there. The purpose of this study was to elucidate the problem of "the formation of races by means of selection of biotypes."[19] But his attempts to elucidate this problem were extended to other objects quite different from insects. His earlier work on variation in the water snail *Lymnaea stagnalis* is, I think, almost completely unknown.

In the early fall of 1923 Dobzhansky and one of his students from

---

therefore that a taste for collecting beetles is some indication of future success in life!" (*The Autobiography of Charles Darwin, 1809–1882*, ed. Nora Barlow, New York: Harcourt, Brace and Co., 1958, p. 63).

[16] For the St. Petersburg response to his work, see Ivan P. Borodin, "Predsmertnoe i bessmertnoe otkrytie S. E. Kushakevicha" [The final, undying discovery of S. E. Kushakevich], *Zhurnal Russkogo Botanicheskogo Obshchestva* 9 (1924). Kushakevich's fame in Petrograd may have had something to do with Filipchenko's invitation to Dobzhansky, Kushakevich's student.

[17] S. Kushakevich, "Popytka tsitologicheskogo obosnovaniia zakonov nasledstvennosti" [An attempt to ground the laws of heredity in cytology], *Priroda*, October 1914, pp. 1205–1220.

[18] For a discussion of the importance of Dobzhansky's entomological training in shaping his approach to evolution, see N. Krementsov's essay in this volume.

[19] F. Dobrzhansky, "Die geographische und individuelle Variabilität von Harmonia axyridis Pallas in ihre Wechselbeziehungen," *Biologische Zentralblatt* vol. 44, no. 8 (1924), pp. 401–426.

Kiev University, L. Kossakovskii, collected snails from different habitats and studied the quantitative variation of shell shape in relation to habitat. When different forms were found they decided to carry out an experiment to corroborate the hypothesis of the heritability of shell shape. This study was carried out by Kossakovskii in the spring of 1924. He obtained progeny from snails with different shell shapes from various habitats and kept young snails in aquariums with uniform constant conditions. Preliminary results were presented at the First All-Russian Hydrological Congress in 1924 and were published the following year under the names of Dobrzhanskii and Kossakovskii. They explained that "the study was carried out with the aim of elucidating the question of the race formation in snails in different but neighboring bodies of water."[20] They concluded that in large bodies of water with more different and changeable conditions, the snail populations included a diverse mixture of biotypes and showed greater variation than those of small, closed bodies of water with relatively constant conditions. Kossakovskii's sudden death in the fall of 1924 interrupted the investigation, and Dobzhansky handed over all the experimental data to Boris Rumiantsev, a Leningrad student who continued the work under Dobzhansky's guidance and published the results in 1928.[21]

In the early and mid 1920s, then, Dobzhansky was already interested not only in race formation in insect species but also in the general problem of adaptive genetic variation and population differentiation. These early studies of populational variation in neighboring but ecologically different habitats foreshadowed his American studies of fifteen years later, but in theoretical language they have nothing in common; in 1924 he described variation as purely morphological and in terms of biotypes and biotype selection.

Although Dobzhansky continued his studies of variation in ladybird beetles in Leningrad, he also began genetic investigations of manifold effects of mutations in *Drosophila*. His large article on this subject is devoid of anything about its evolutionary implications.[22]

[20] *Trudy I Vserossiiskogo Gidrologicheskogo S"ezda* [Proceedings of the first all-Russian hydrological congress], 1925, pp. 481–482.

[21] B. Rumiantsev, in *Trudy Leningradskogo Obshchestva Estestvoispytatelei* vol. 58, no 2 (1988), pp. 45–64. The article describes Kossakovskii's experiments; Dobzhansky's supervision of the work is discussed in his correspondence with Filipchenko.

[22] Th. Dobzhansky, "Studies on manifold effects of certain genes in *Drosophila melanogaster*," *Zeitschrift für induktive Abstammungs- und Vererbungslehre* vol 43 (1927), pp. 330–388.

But the work began in Kiev in 1923 with his first experiments on the phenogenetics of the sexual apparatus of *Drosophila*. As a young enthusiastic Russian scientist, he had already "let the cat out of the bag," by revealing the evolutionary significance of his studies. He began a very brief preliminary report of this study with these words: "As we know, different species of insects, as well as geographical races of the same species, differ in the organization of their sexual apparatus. Those differences have a great biological significance for their bearers because they make mating of differently organized forms difficult and can cause isolation."[23] In this particular study, as well as later ones, Dobzhansky examined the morphology of sexual apparatus (mostly female) in different mutations of *Drosophila melanogaster* and found that mutations which were described by changes in eye shape, eye color, body color, and other external characteristics also influenced the morphology of internal sexual organs.

At roughly the same time, Dobzhansky was studying the morphology of the female sexual apparatus in ladybird beetles and showing its significance for distinguishing not only species but also genera, tribes, and subfamilies.[24] The morphology of the sexual apparatus, then, provided opportunities to investigate a series of interconnected problems: the appearance of differences through mutation; racial and species differences; and the origin of macroevolutionary patterns—differences between genera and higher taxa.

In the mid-1920s, evolution for Dobzhansky was still a historical process of morphological changes, not changes in gene frequencies or in the genes themselves. Phenogenetic analysis of gene manifestation and gene effects on the formation of characteristics seemed to be a necessary link between the theory of heredity and the theory of

[23] Th. Dobzhansky, "Über dem Bau der Geschlechtsapparats einiger Mutanten von Drosophila melanogaster," *Zeitschrift für induktive Abstammungs- und Vererbungslehre* vol. 34 (1924), p. 245. In his memoirs Dobzhansky indicated that the publication of this report attracted Filipchenko's attention and led to Dobzhansky's invitation to Leningrad. But this paper was received by the journal on 20 September 1923 and published in early 1924, after Dobzhansky had already arrived in Leningrad. Moreover, the published version of the paper indicated that it was "von Th. Dobzhansky, Leningrad," which means that the journal had been informed of Dobzhansky's position before publication. I want to stress that at the time Filipchenko invited him to Leningrad, Dobzhansky had published nothing whatever on variations or genetics, so it must have been based on personal or informal information.

[24] F. G. Dobrzhanskii, "Polovaia sistema bozh'ikh korovok (Coccinellidae) kak vidovoi i gruppovoi priznak" [The reproductive system of ladybird beetles as a species and group character], *Izvestiia Akademii Nauk SSSR* (1926), pp. 1385–1394 and 1556–1586.

evolution. At that time Dobzhansky was more interested in characteristics than in genes or mutations; for him genes and mutations
were only a means to understand the appearance and evolution of
new forms.

Dobzhansky's early approach to evolution is illuminated by two
little-known articles he wrote in the mid-1920s and published in
popular journals—articles usually omitted from his published bibli-
ographies. The first, "Inheritance and mutation" (1924), popularized
Morgan's genetics. But Dobzhansky's relatively straightforward description of the chromosomal theory was followed by a statement of
the evolutionary significance of genetics that was entirely his own.
"The significance of mutations for evolution," he wrote, "consists in
this, that their characteristics, combining with one another (according to Mendel's laws), become the material for the action of natural
selection.... The continuous shuffling of genes occurs everywhere
under our eyes and so there appears an enormous quantity of gene
combinations; it can easily occur that some of the combinations
turn out to be harmonious, better, more adapted to life than others.
Due to the action of natural selection some of these combinations will become extinct, others will turn out to be neutral, and still
others useful. Immediately after their appearance, new mutations
will fall into this sorting apparatus of recombination and selection, and as soon as a harmonious combination is created it will
be consolidated in nature and a stage of evolution will be accomplished."[25]

The other paper, entitled "Mutations and speciation" (1926), dealt
with evidence of the role of mutations in speciation and evolution in
general.[26] It was an essay review of a kind, and here Dobzhansky included nothing at all about mutual isolation of species or about mutations or morphology of the sexual apparatus. Although he did not
subsequently cite his own article, there is a striking similarity between passages in this article and the second chapter of his 1937
book, *Genetics and the Origin of Species*. For example, in order to
show that mutations are natural and may play a role in evolution,
Dobzhansky presented many examples in his 1926 article of similar-

[25] F. G. Dobrzhanskii, "Nasledstvennost' i mutatsiia" [Inheritance and mutation],
Chelovek i Priroda, 1924, no. 5/6, pp. 417–426; passage translated into English by Daniel P. Todes.

[26] F. G. Dobrzhanskii, "Mutatsii i vidoobrazovanie" [Mutations and speciation],
Priroda, 1926, no. 5/6, pp. 31–44.

ity between mutations (mainly in *Drosophila* but also in other organisms) and specific characteristics of certain species, such as wingless flies. Most of these examples reappeared eleven years later in his book. His discussion of the problem of the viability of mutations in 1926 and 1937 was also very similar, the only difference being the experimental evidence he cited: in 1926 he cited Pearl, in 1937 Timoféeff-Ressovsky.[27]

Dobzhansky concluded his article with a gradualist description of the process of speciation. "The process of mutation," he wrote, "introduces the hereditary differences which, summed up through the activity of natural selection, provide the beginning of species distinctions. Different species, even those that are very close, are always distinguished from one another by a great quantity of various hereditary factors. Furthermore, a factorial mutation, however sharp its characteristics, however great the taxonomic value of those characteristics, nevertheless leads to the formation of a form that differs from the initial form by only one, single factor. It is important, nevertheless, that by means of selection from mixed material there can be created various forms differing from one another by any number of genes. This very process of the gradual differentiation of a mixed population, proceeding in nature, is the process of speciation."[28]

During his Russian period, then, Dobzhansky expounded a more or less complete picture of his contemporary views on evolution and speciation and carried out a series of investigations on the problem of race and species formation. However, I would stress that this was Dobzhansky *before* the evolutionary synthesis. He understood the role of isolation and had even attempted to understand it through studies of the phenogenetics of the sexual apparatus, but he did not yet have a populational concept of speciation. It was still not mutations but "characteristics combining with one another"—not populations but "forms" differing from one another by "any number of genes."

[27] Although the 1926 articles include no references, it is quite possible to find his sources. During his stay in Leningrad, Dobzhansky intensively reviewed foreign genetic literature and published several short reviews in *Priroda* (e.g., "Eksperimental'noe izuchenie dlitel'nosti zhizni" [Experimental study of longevity], 1925, no. 10/12, pp. 115–116, which deals with Pearl's experiments). Except for Filipchenko, Dobzhansky was the only Leningrad geneticist who published reviews and articles in this distinguished journal.

[28] Dobrzhanskii, "Mutatsii i vidoobrazovanie," p. 44 (passage translated by Daniel P. Todes).

Conclusion

Dobzhansky often praised Filipchenko as his mentor in genetics but disparaged his teacher's evolutionary views. In retrospect, we can see their differences as deriving in part from their different training. Filipchenko's initial training in comparative anatomy and embryology led him to be concerned with macroevolutionary questions. Dobzhansky's initial training, by contrast, was in entomology, collecting and classifying ladybird beetles with their luxuriant adaptive variation, and this centered his concern on microevolution. The two men's backgrounds also influenced their views on speciation. Filipchenko's studies of inheritance involved species crosses, so he was certain that different species could interbreed. By contrast, Dobzhansky studied the significance of differences in sexual apparatus for species discrimination, a concept of mechanical isolation common among entomologists.[29]

For all their differences, however, Dobzhansky's and Filipchenko's views in the 1920s had many similarities. Both men were trained zoologists committed both to evolution and to the new experimental biology. Both saw Mendelian genetics and biometrics as complementary. Both understood evolutionary variation and speciation not in terms of mutations but in terms of morphological characteristics and forms. And for the reason that the relation of genetics to evolution depended on the phenotype, both regarded studies of phenogenetics, of the gene's manifestation and its influence on development, as a significant part of the experimental study of evolution.[30] The very same approach characterized the work of Sergei Chetverikov and his school: for them, as for Dobzhansky, phenogenetical studies and studies of variation within natural populations were two sides of the same coin.[31] Only in the late 1930s, when "population genetics" came to be regarded as the only way to study the genetic

[29] See N. Krementsov in this volume. For an excellent analysis of the role of scientific training, see Garland E. Allen, "Naturalists and Experimentalists," *Studies in History of Biology*, ed. W. Coleman and C. Limoges (Baltimore: The Johns Hopkins University Press, 1979), pp. 179–209.

[30] For an example of Dobzhansky's continuing work in phenogenetics, see Th. Dobzhansky and F. N. Duncan, "Genes that affect early developmental stages of *Drosophila melanogaster*," *Roux Archive* vol. 130, no. 1 (1933), pp. 109–130, which refers explicitly to Filipchenko's studies in phenogenetics.

[31] On Chetverikov and phenogenetics, see Mark B. Adams, "The Founding of Population Genetics: Contributions of the Chetverikov School, 1924–1934," *Journal of the History of Biology* vol. 1, no. 1 (1968), pp. 23–39; and Mark B. Adams, "Chetverikov, Sergei Sergeevich," in *Dictionary of Scientific Biography* vol. 17, supp. 2, ed. Frederic L.

basis of evolution, was the phenogenetic, developmental approach largely abandoned.[32]

The absence of evolutionary references in Dobzhansky's genetic publications on *Drosophila melanogaster* between 1924 and 1936 can make it appear that his 1937 book came "out of the blue." The observation is accurate enough; "evolutionary metaphysics" is absent from his earliest articles on the manifold effects of mutations and from virtually all his subsequent genetic papers on translocations, hybrid sterility, and so forth through 1936. But we should note that most of these articles were published in English. In writing them, Dobzhansky may have been following the canons and conventions of laboratory research reports; he may even have been consciously mimicking the tightly analytical style of papers published by the Morgan group.

Dobzhansky's Russian writings, however, reveal another story. In his Russian review of studies on translocations, he pointed out that these studies are interesting and significant precisely because of their evolutionary implications.[33] And in his letters to Filipchenko, written from the United States between 1928 and 1930, he explained every step in his genetic research in terms of its significance for evolution. As we investigate Dobzhansky's Russian writing and background more thoroughly, we may come to a much richer and more complete picture of the development of the evolutionary genetics of Iurii Filipchenko's most distinguished and influential protégé.

Acknowledgments

I am grateful to many people and institutions for help and support. My special thanks go to Daniel Todes who provided encouragement in Leningrad and who helped me with translating and editing my

---

Holmes (New York: Charles Scribner's Sons, 1990), pp. 155–165. In Russian, see V. Babkov, *Moskovskaia shkola evoliutsionnoi genetiki* [The Moscow school of evolutionary genetics] (Moscow: Nauka, 1985). In his detailed study, Babkov declared the phenogenetic approach to be an original contribution and central theme of Chetverikov's research program. But as we see, this approach was widely shared by many other zoologists who converted to genetics; in the case of Chetverikov (as well as Filipchenko and Dobzhansky) it is due not to his originality but to the nature of the transition from zoological to genetical practice.

[32] For a discussion of this, see Mark B. Adams, "La génétique des populations était-elle une génétique évolutive?" *Histoire de la Génétique: Pratiques, Techniques et Théories*, ed. Jean-Louis Fischer and W. H. Schneider (Paris: A.R.P.E.M., 1990), pp. 153–171.

[33] *Trudy po prikladnoi botanike, genetike i selektsii* vol. 6 (1934), pp. 147–171.

first draft. To Garland Allen and Dick Burian, whose discussions and papers I found very helpful, I owe much. Mikhail Konashev kindly provided me with his typescript of the Dobzhansky-Filipchenko correspondence, which was very helpful. But most of all, I am indebted to Mark Adams for his encouragement and support, for the special care with which he guided me and this paper through to the final draft, and for his enormous work in editing my text.

From the Archives: Dobzhansky in Kiev and Leningrad

Mikhail B. Konashev

IN WRITING about Soviet genetics in the 1920s, Theodosius Dobzhansky rightly focused on three great centers: Nikolai Vavilov's Institute of Applied Botany, Nikolai Kol'tsov's Institute of Experimental Biology in Moscow (which included S. S. Chetverikov and A. S. Serebrovskii), and Iurii Filipchenko's department of genetics at Leningrad University (Dobzhansky 1980, p. 236).

Of the three centers, Filipchenko's has received by far the least attention, especially by Russian historians. There are several Western works devoted to Filipchenko and particularly to his eugenics (Adams 1980a, 1980b, 1989, 1990a, 1990b), but in Soviet publications, aside from a slim and greatly censored biography (Medvedev 1978), Filipchenko is usually mentioned only as a scientist who determined the "historical fate" of the genetics department of Leningrad University (e.g., Prokof'eva-Bel'govskaia 1978, p. 3). In such accounts, of course, almost no mention is made of Dobzhansky or his role.

On the basis of research in Russian and Ukrainian archives, I have been able to establish many facts about Dobzhansky's life in Kiev, his connections with Filipchenko, and his move to Leningrad and the role he played there. In some sense, the collaboration and friendship of these two men had a great influence on the development of genetics in Leningrad and in Russia as a whole.

DOBZHANSKY IN KIEV

Theodosius Dobzhansky was born in 1900 in Nemirov, a small town 200 kilometers southwest of Kiev. His father, Grigori Dobrzhanskii, was a mathematics teacher in the town's gymnasium. In his unpublished autobiographical reminiscences, Dobzhansky stated that all the male ancestors on his mother's side that he had ever heard about were Orthodox priests. He described one relative as "a priest in some town in the province of Kiev whose name I don't know" (Dobzhansky, 1962–1963, p. 4).

Fig. 1. Fedor Petrovich Dobrzhanskii, archpriest, 1914.
From the Central Russian Photoarchive, fund 175, no. 1429, item 7.

It is possible that this man was Fedor Petrovich Dobrzhanskii, archpriest (*protoirei*) and teacher of religion (*zakonouchitel'*—literally "teacher of God's law") in the Nemirov gymnasium. He was born in May 1857 in Zelentsal, a village in Podolskaia Province where Nemirov was situated. In a letter dated 5 February 1913 Fedor wrote that his father and son were priests, his mother was a daughter of a priest, and in general all his kin on both his father's and mother's sides were priests.[1] The fate of these various priests is obscure. Dobzhansky's religious pedigree is worth further study as it may cast light on his subsequent philosophical and religious views.

In 1910 the Dobzhanskys moved to the outskirts of Kiev where they lived in a small one-floor house (Ossievskaia Street 23, apartment 4). Now it is called Hertsen Street. A part of old Ossievskaia Street is

[1] TsGB ANUk, fund 175, delo 1429, p. 3. *Editor's note:* Russian archival sources are listed by *fond* ("fund"), *opis'* ("box," "shelf," or "subdivision"), *delo* ("folder," "file," or "dossier"), and *list* ("page"). In the notes, *fond* and *list* are given as *fund* and *page*, whereas *opis'* and *delo*, which do not have an adequate English rendering, are left in transliterated Russian. See the references for other archival abbreviations.

preserved, including some houses of the kind Dobzhansky lived in, but the Dobzhansky house no longer exists.[2]

Early Professional Life

The Dobzhansky family spent the troubled Civil War years in and around Kiev. In his oral memoirs Dobzhansky vividly recounts the repeated taking and retaking of the city by various opposing armies.

Dobzhansky's principal job was as an assistant at the Kiev Agricultural Institute (until 1923 the agricultural faculty of the Kiev Polytechnical Institute). After the death of his mother in 1920, he moved into a new residence (Politekhnikum, no. 3, apartment 38).[3] Like other assistants and professors of the Kiev Polytechnical Institute, Dobzhansky lived in institute housing on its grounds. The address was quite near Kiev University.

The lists of Dobzhansky's positions given in Russian articles have been incomplete.[4] The most complete list thus far, written by Dobzhansky and found in a Kiev archive,[5] is as follows.

February 1918–September 1919: assistant of academician V. I. Vernadskii in the Ukrainian Academy of Sciences

1919: teaching in "Zero" semester at Kiev University

1919–1920: assistant at the Odessa Agricultural Institute

April 1919–1923: zoologist at the Zoological Museum of Ukrainian Academy of Sciences

April 1919–1920: lab assistant at the Kiev Agricultural Institute [agricultural faculty, Kiev Polytechnical Institute]

24 April 1920–1923: assistant at the Kiev Agricultural Institute [agricultural faculty, Kiev Polytechnical Institute]

September 1922–1923: lecturer at the Kiev Agricultural Institute [agricultural faculty, Kiev Polytechnical Institute]

We should note that during this period Kiev hosted many people who were (or would become) prominent in the history of science, and Dobzhansky was in contact with them. In addition to his connections with V. I. Vernadskii,[6] he also knew Ivan I. Schmalhausen, then at Kiev University, and the two often liked to stroll together on

---

[2] Old houses at the end of this street were demolished, and a large new building of the Kiev Institute of Political Science (property of the Communist Party of Ukraine) was built on the site.

[3] GAK, fund 179, opis' 2, delo 24, p. 1.

[4] Naumov 1989, p. 1131; Gall and Konashev 1990, p. 80.

[5] GAK, fund 176, opis' 2, delo 24, pp. 1, 2, and 9.

[6] See Kendall E. Bailes, *Science and Russian Culture in an Age of Revolution. V. I.*

Bibikovskii Boulevard (now Taras Shevchenko Boulevard). In addition, Dobzhansky shared his apartment with N. Wagner and G. A. Levitskii, the latter a leading cytologist who published the first Russian textbook of cytogenetics at precisely the time Dobzhansky was living with him (Adams 1990c).

Like most people during these difficult times, Dobzhansky held down several other jobs and was constantly forced to search for additional work to supplement his income. There are several curious pages in Dobzhansky's personnel dossier in the Kiev archives. One is headed "Form five. Kiev Regional State Archive of Department of the Ministry for Internal Affairs. List of employment documents."[7] The eighth page in the same file is an official form for the Kiev Polytechnic Institute dated 17 January 1922. One item on the form says: "List all of your work experiences (in factories)," to which Dobzhansky responded: "Factories are not my specialty."[8] But he did lecture on zoology and general biology at the Polytechnical Institute's "Rabfak" (courses for young working people) and on biology at the Kiev Pedagogical Institute.[9]

It was in Kiev that Dobzhansky's experience with scientific fieldwork actually began. His first official expedition was to investigate pests of the sugar beet in Kiev Province, which he organized and carried out in 1922.[10] In May 1923 he carried out two additional expeditions: the first to Melitopol in the Tauride region of the Crimea,[11] the second from May 15 through 29 to Askania-Nova to investigate "bezkhrebtovie" (invertebrates).[12]

Discovering Genetics

In 1921, after the Civil War had ended, scientific literature began to arrive in Kiev and Dobzhansky read Filipchenko's article "Chromo-

---

Vernadsky and His Scientific School, 1863–1945 (Bloomington: Indiana University Press, 1990).

[7] GAK, fund 176, opis' 2, delo 24, unnumbered page.

[8] GAK, fund 176, opis' 2, delo 24, unnumbered page.

[9] GAK, fund 176, opis' 2, delo 24, pp. 1, 2, and 9; PA RAN, fund 132, opis' 1, delo 45, p. 251.

[10] GAK, fund 176, opis' 2, delo 24, p. 4.

[11] PA RAN, fund 291, opis' 2, delo 52, p. 2 The exact purposes of the expedition to Melitopol' are unclear from the documents. [*Editor's note.* See Krementsov's article in this volume for more information about this.]

[12] GAK, fund 176, opis' 2, delo 24, p 29.

somes and Heredity," which had been published in the popular science magazine *Priroda* in 1919 and was devoted mainly to the investigations of the Morgan school. The article began with a short history of ideas of heredity from the 1880s, then expounded the results of T. H. Morgan and his group. In this review Filipchenko used all the most important articles of the members of the Morgan group, but chiefly their book *The Mechanism of Mendelian Heredity* (Morgan et al. 1915). Several figures from the book were reproduced verbatim.

Filipchenko published a second article on the works of the Morgan school in 1922 in the same journal. Drawing on the 1921 German edition of Morgan's classic, Filipchenko argued that the works of the Morgan school had given genetics the status of an exact or "positive" science: the data had been arrived at by experimental methods and had established exact laws comparable to any other positive science and on a par with physics or chemistry (pp. 51–53, 65). Taken together, Filipchenko's two articles summarized the work of the Morgan school in sufficient detail for a reader to understand genetics, appreciate its achievements and significance, and even begin research work.

Dobzhansky, who had read Darwin's *Origin of Species* at age fifteen, was much impressed by the articles. "To me these reviews were a revelation. To most senior biologists *Drosophila* mutants were a collection of monstrosities, of no significance for evolution" (Dobzhansky 1980, p. 232). The articles contained rich descriptions and analyses of *Drosophila* mutations, which led Dobzhansky to see that the principles of genetics and evolutionary theory were compatible.

The impact of reading Filipchenko's articles reinforced things happening in Dobzhansky's life in Kiev. At the time, Dobzhansky shared an apartment with cytogeneticist G. A. Levitskii on the grounds of the Kiev Polytechnical Institute where they both worked. In the winter of 1921–1922, Levitskii spent about two weeks working in Nikolai Vavilov's library in Petrograd, which was one of the few repositories for the new literature that was finally coming into Russia from abroad. Upon his return to Kiev, Levitskii shared his knowledge and enthusiasm over the new findings in his course lectures and also in long discussions over the kitchen table at home. As Dobzhansky later remarked, when Morgan's genetics "was being discussed in the kitchen, ... it became clear to me that is what I must be doing" (1962–1963, p. 119).

Fig. 2. Dobzhansky in the early 1920s.
Photo courtesy of Y. L. Goroshchenko.

However, Dobzhansky did not regard Kiev as the ideal location for doing *Drosophila* genetics. He was one of a very few geneticists in Kiev and would have had to work basically alone. Dobzhansky had experience working in groups and liked to work in "friendly collectives" (*tovarishchestvo, artel'*), so the availability of a group of other geneticists was important to him, and he found the groups that were forming in Petrograd and Moscow attractive (Konashev 1991a, pp. 61–62).

On the Move

Beginning in 1921, Dobzhansky's letters reveal an interest in relocating. In a letter to V. I. Vernadskii dated 7 September 1921, he wrote: "The vegetation in Kiev is so boring, and my complete solitude prompts me to long for a change of setting."[13] A year later, on 22 February 1922, he again wrote to Vernadskii: "Now I have absolutely not the slightest moral connections with Kiev; as to financial ones I am ready to sacrifice everything, and somehow set myself up."[14] Finally, in writing to Bial'nitskii-Berul'ia in Petrograd, he indicated that he was visiting there "with the purpose of working and maybe with the purpose of setting myself up at a place of residence."[15] And, of course, Dobzhansky wanted to study *Drosophila* genetics.

Dobzhansky made the necessary applications for several trips to Petrograd and Moscow in 1922 and 1923. The first time he asked a dean of the agricultural faculty of Kiev Polytechnical Institute to send him to Petrograd to work at the Zoological Museum and Library of the Academy of Sciences for a month, from 25 July through 25 August. In his second application Dobzhansky wrote that he was going to take part in the Fourth All-Russian Entomology-Phytopathology Congress in Moscow and in the First All-Russian Zoological Congress in Petrograd 6–26 December 1922.[16] In a letter dated 23 April 1923 to A. A. Bial'nitskii-Berul'ia, who was on the staff of the Zoological Museum, Dobzhansky wrote of his intention to be in Petrograd in late summer or autumn 1923.[17] He received the necessary documents for his visit to Petrograd and Moscow in late July 1922, including documents to stay at the Petrograd House of Scientists, a residence for visiting scholars.[18] He also received permission and the necessary documents for his attendance at the meetings and took part in them.[19]

In his memoirs Dobzhansky commented on these trips. In the Houses of Scientists, he recalled, "you got acquainted with your colleagues in other cities, of course, namely in Moscow and Petrograd. I met and became acquainted with biologists in both places, some of

[13] Sorokina 1990, p. 93.
[14] Ibid., p. 94.
[15] PA RAN, fund 291, opis' 2, delo 52, p. 1.
[16] GAK, fund 176, opis' 2, delo 24, pp. 21–22.
[17] PA RAN, fund 291, opis' 2, delo 52, p. 1.
[18] GAK, fund 176, opis' 2, delo 24, pp. 18–19.
[19] GAK, fund 176, opis' 2, delo 24, pp. 23–24.

whom were to become my good friends" (Dobzhansky 1962–1963, p. 122). During one of his visits to the Kol'tsov Institute in Moscow, he "met a number of people there, among them people quite important in the history of genetics," namely Chetverikov and Serebrovskii.

Dobzhansky's trips to Petrograd and Moscow apparently had a great effect on him. He returned from Petrograd convinced that he had to learn English. In the gymnasium he had learned French and German, but he discovered in Petrograd that a lot of the foreign literature was in English. He took several English lessons and then tried to learn the language with the aid of a scientific book and a dictionary. Before he left Russia in 1927, he had already started to understand English well, as is clear from some of his articles during the period. After only one month in the States, he could speak English relatively fluently.

Dobzhansky's trip to Moscow had a more immediate scientific result. In Chetverikov's laboratory Dobzhansky took a collection of mutant strains of *Drosophila melanogaster* and started to work on them enthusiastically. He examined in detail the genitalia of mutants and found that they had statistically significant differences in various parts of their reproductive apparatus, particularly in the shape of the spermatheca. Thereby he demonstrated that a change in one gene can modify different traits. These results were first reported to Dobzhansky's colleagues and then published (Dobzhansky 1924a). Filipchenko was interested in Dobzhansky's work on this and sent him an offer to join his department in Petrograd. Dobzhansky agreed.

The offer must have been most welcome. Dobzhansky's career in Kiev did not appear promising. In 1923 a new veterinary institute was being organized in Kiev, and there was to be a post for a zoology professor. Dobzhansky was one of the two candidates for this professorship; the other was Krasheninnikov (a nephew of Prof. S. E. Kushakevich). But Dobzhansky made one slip during his interview with the director of the institute, Prof. Omelchenko, a leading Ukrainian nationalist. Dobzhansky didn't show himself to be a great Ukrainian patriot and was not selected.[20] In addition, Dobzhansky wanted to leave Kiev and to be able to work in a community of geneticists. Thus, Filipchenko's invitation to join him at the country's premier university, in its cultural capital, was just what Dobzhansky had wished for.

[20] Dobzhansky 1962–1963: 143–145.

LENINGRAD

Years later, in accounting for why Filipchenko had picked him (e.g., in his oral memoirs), Dobzhansky seemed both perplexed and grateful. But he was being too modest. By 1924 Dobzhansky was a well-connected young man who had good ties with the Zoological Museum and had published and conducted fieldwork in entomology (Filipchenko's original love). This same young man had excellent university training in the Kiev school of cytogenetics (under Kushakevich) and extensive teaching, laboratory, and field experience. Furthermore, he had experience in practical agricultural settings and had organized three expeditions. Whether or not they had ever met, Dobzhansky had acquired an ideal set of qualifications for the job Filipchenko wanted done.

Filipchenko's School

In Dobzhansky's retrospective judgment, Filipchenko "built a school perhaps smaller than those of Chetverikov and Serebrovsky, but his impact on the evolutionary and genetical thought in Russia was, if anything, greater" (Dobzhansky 1980, p. 236). The reason was that Filipchenko, perhaps more than any other contemporary, was responsible for the establishment of genetics as a discipline in Russia. The author of seven original textbooks in the early and mid 1920s, Filipchenko almost single-handedly created the basic Russian literature in the field (Adams 1990b).

Thus, Filipchenko's "school" did not focus on any particular organism or method but rather laid out the field's territory and scope in all its breadth: eugenics, the genetics of domesticated animals, plant genetics (principally wheat), and Morganist *Drosophila* genetics. Filipchenko's group was less a school than an amalgam of three interrelated organizations that he had created.

The first was his department at Petrograd (after 1924, Leningrad) University. In 1918 Filipchenko became professor at Petrograd University and created its laboratory of genetics and experimental zoology. In 1919 this laboratory became the university's Department of Genetics and Experimental Zoology (Medvedev 1978, pp. 12–13).

The second was Filipchenko's Laboratory of Genetics and Experimental Zoology, created in 1920 outside the city on the Bay of Finland, at the Peterhof Scientific Institute, of which he was the scientific secretary.

The third was the Bureau of Eugenics of the Commission for the Study of Natural Productive Forces (KEPS) of the Russian Academy of Sciences, founded on Filipchenko's initiative in February 1921. Originally it was based at Filipchenko's university department; subsequently it was given quarters a few blocks away. Although its name was changed (1925, Bureau of Genetics and Eugenics; 1927, Bureau of Genetics; 1930, Laboratory of Genetics), its staff and basic research profile remained more or less fixed throughout the 1920s.

Taken together, these three organizations—the department, the Peterhof laboratory, and the bureau—were sufficient to support scientific research and the livelihood of the researchers. Filipchenko's whole group used the equipment, space, apartments, and financial resources of all three organizations more or less interchangeably, depending on the object of research. Eugenics work took place in the bureau; plant genetics at Peterhof; domesticated animal work in the bureau and on its expeditions; *Drosophila* genetics principally at the university. But members of the group participated in more than one area, some in all areas.

The staff of Filipchenko's shop was relatively young. In 1923 Filipchenko was forty-one. One of his two chief assistants was D. M. D'iakonov, who died in September 1923. The other was V. M. Isaev. Isaev had returned from World War I to Petrograd University in 1918 but had received permission to teach only in November 1921.[21] In the summer of 1924, at the age of thirty-five, he died suddenly during a mountain-climbing vacation in the Caucasus.[22] A. I. Zuitin, Ia. Ia. Lus, and T. K. Lepin, three students from the Baltic states, were confirmed by the board of Petrograd University as scientific workers in Filipchenko's department for a term from 1 October 1923 to 1 January 1925. At the time of their appointment, Zuitin was thirty-two, Lepin twenty-eight, and Lus twenty-six.[23] Two other assistants of Filipchenko were K. A. Adrainova-Fermor and a technical assistant, I. E. Bordzio.[24]

The group's base was its university quarters. In early 1922 the genetics department moved from a building on the 16th Line to the

[21] TsGA SPb, fund 7240, opis' 14, delo 144, p. 10r.

[22] TsGA SPb, fund 7240, opis' 14, delo 154, pp. 50, 77, 137r.

[23] TsGA SPb, fund 7240, opis' 14, delo 154, p. 49

[24] *Nauchnye uchrezhdeniia*, p. 148.

building of the former History and Philosophy Institute at 7 University Embankment (now occupied by the philology faculty of St. Petersburg University).[25] The department was located on the first floor of the building and consisted of Filipchenko's office, a corridor, a lecture hall, and three rooms for researches. Most of the department's windows faced the Neva River, St. Isaac's Cathedral, Senate Square, and the Bronze Horseman.[26]

From the beginning, Filipchenko himself taught most of the courses in the department: "Introduction to Genetics," "Experimental Zoology," "Biometry," "Evolutionary Theories," and "Special Course in Genetics."[27] Isaev had taught "Cytological Basis of Heredity," a course taken over by cytologist I. I. Sokolov in 1926–1927. Two unsalaried docents gave "applied" courses: V. P. Nikitin ("Zootekhnika," animal breeding) and B. E. Pisarev ("Selection").[28] In 1922 there were only two students in the department, N. N. Medvedev and N. Ia. Fedorova; in 1923–1924 there were already twenty (Medvedev 1978, p. 13).

Dobzhansky in Leningrad

Dobzhansky arrived in Petrograd the night Lenin died. His first quarters in the city (which was renamed Leningrad the next day) was a flat at 21 Dobroliubova Street, a five-minute walk from Leningrad University (first alone, then with his wife, Natasha). As it turned out, the lady from whom they rented the rooms was a prosperous, attractive prostitute. Soon Natasha demanded that they find another flat. In 1925 they moved to Bolshoi Prospect 44, apartment 29, in the same Petrogradskii district of the city, a twenty- or thirty-minute walk from the university and on a tram line.

This was the flat where the Dobzhanskys lived until their departure for the United States in December 1927. Its entrance was off the courtyard from 2 Oranienbaumskaia Street. They rented two rooms in their landlady's apartment: a bedroom (overlooking the courtyard) and a living room (which also served as a study) whose two

---

[25] TsGA SPb, fund 7240, opis' 14, delo 142, p. 211r; delo 144, p. 18r.

[26] For more information about the physical setup, see Medvedev 1978; and Konashev 1993b, in a forthcoming volume on the history of genetics at St. Petersburg University, a principal theme of which is Dobzhansky's contribution to that history.

[27] TsGA SPb, fund 7240, opis' 14, delo 193, pp. 67–67r.

[28] Ibid., pp. 66–67.

Fig. 3. Department of Genetics, Leningrad University (winter holiday, 1926–1927). *Bottom row, left to right*: E. P. Gogeizel (then Radzabli), Iu. A. Filipchenko, N P. Sivertseva (already Dobzhansky's wife); *Top row*: Dobzhansky with almost all of his students: *left to right*, Ia. Ia. Kerkis, G. I. Shpet, S. Ivanov, Dobzhansky, T. K. Lepin, Y. L. Goroshchenko, and M. M. Levit. Photo courtesy of Y. L. Goroshchenko.

windows overlooked Strelninskaia Street. The living-room furniture consisted of a sofa, dining table, sideboard, and writing table. This landlady, however, was an older woman who was extremely rigorous about noise; this created problems, because Dobzhansky was gregarious and liked company. As a result, the Dobzhanskys often visited the nearby apartment of his senior student, Iulii Kerkis (Kronverkskaia Street, 1, also in Petrogradskii district). Kerkis lived with his parents, and his mother was a good cook, so the Dobzhanskys often dined there.

Soon after arriving in Petrograd, Dobzhansky became Filipchenko's right-hand man in various scientific enterprises. Dobzhansky was officially confirmed by the university board as an assistant in the department of genetics as of 13 December 1923 (more or less as a replacement for D'iakonov).[29] Following Isaev's death in 1924, Dob-

[29] Ibid., delo 152, p. 147

Fig. 4. Filipchenko's group on summer research at Staryi Peterhoff
(on the Bay of Finland, near Leningrad), mid 1920s.
Top row, left to right: I. F. Bordzilo, A. I. Zuitin, T. K. Lepin, Iu. A. Filipchenko,
Dobzhansky, V. I. Savel'ev, I. I. Kanaev, G. M. Pkhakadze. *Bottom row,
left to right*: N. N. Medvedev, Egorova, N. P. Sivertseva(-Dobrzhanskaia),
Ia. Ia. Lus, E. P. Gogeizel (Radzabli), R. A. Mazing, N. Ia. Fedorova.
Photo courtesy of Y. L. Goroshchenko.

zhansky became director of departmental property. Officially he became a staff member of the bureau only on 1 December 1925, just
after it had been renamed the Bureau of Genetics and Eugenics.[30]

At the university department, Dobzhansky (together with his student Iulii Kerkis, who had joined him from Kiev) worked in one of
the three research rooms that overlooked the courtyard. His own researches focused almost exclusively on ladybird beetles and *Drosophila* genetics. He didn't share Filipchenko's passion for eugenics
and zootechnics; later, from the States, Dobzhansky wrote to Filipchenko urging him to "keep this zootechnical business at arm's
length."[31] The one exception was the work Dobzhansky carried out
on domesticated animals as head of the KEPS expeditions of the Bureau of Genetics and Eugenics in the summers of 1926 and 1927.

[30] PA RAN, fund 132, opis' 1, delo 45, pp. 106–106r.
[31] NRB, fund 813, delo 283, p. 16.

The KEPS Expeditions

Departmental lore has it that the bureau's summer expeditions to Central Asia were cooked up by Filipchenko's younger collaborators, including Dobzhansky, so that they could enjoy exotic travel and pursue their own research interests. Dobzhansky was very interested in collecting ladybird beetles in the Tian Shan mountains. He and his wife spent a delayed honeymoon in the summer of 1925 in Central Asia, chiefly in East Fergana and Central Tian Shan, where he began to collect ladybird beetles.[32] Of course, there were other reasons as well for the expeditions: Filipchenko was interested in animal breeding and breeding stock, and KEPS was mounting enormous, broadranging expeditions to the republics of Central Asia. And, of course, the younger workers needed summer support.

Dobzhansky headed the bureau's expeditions to the Kazakhstan region during the summers of 1926 and 1927. His age created some problems. In the diary that he kept on his first expedition to Kazakhstan, for example, Dobzhansky wrote that when he presented the research program to the Commissariat of Agriculture on 28–29 May 1926, he was greeted with a certain distrust which he attributed to his young age;[33] only when he visited the commissar at home later that evening was the matter finally settled. Under 15 May in his diary Dobzhansky complained that Lus had accused him of using the expedition to pursue his own interests.[34] Probably the age difference (Dobzhansky, the head of the expedition, was twenty-six, Lus twenty-nine), combined with the fact that a newcomer had been promoted over those with seniority, played some role in these and other conflicts on the expeditions.

The trips were physically difficult, but Dobzhansky did not seem to mind. In his field diary, for example, he wrote under 28 May 1926 that not all members of the expedition were contented with the food (milk, tea, and bread) but that he did not feel any adverse consequences from this diet. The result of these expeditions was subsequently published (Dobzhansky 1927a, 1928). They were clearly memorable events in Dobzhansky's life.

Nonetheless, Dobzhansky decided not to take part in any future expeditions. He regarded the time-consuming preparation and anal-

[32] NRB, fund 813, delo 282, pp. 1–17.
[33] *Dnevnik*, 1926c, p. 8.
[34] Ibid., p. 25.

ysis required by such expedition work as taking valuable time away from his work on *Drosophila*, and in retrospect, he did not regard these expeditions as having had any significant practical value (Dobzhansky 1962–1963, p. 220). He also thought that such work involved "taking other people's chestnuts out of the fire."[35] Thereafter, Ia. Ia. Lus took Dobzhansky's place as head of the expeditions, which continued until 1935.

The Best Laid Plans

Aside from expedition work, Dobzhansky's official research in the bureau dealt entirely with insects. He investigated local and geographical variation in the color and spot pattern of two Coccinellidae genera, *Harmonia* and *Adalia*, and explained the intra- and interpopulational genetic variation as resulting from the same fundamental evolutionary processes (Dobzhansky 1924b, 1924c). His results were given in the report of the bureau's scientific work.[36] Concurrently he continued his *Drosophila* investigations on manifold gene effects (Dobzhansky 1927b). It was this work that led most directly to Filipchenko's nomination of Dobzhansky to the International Education Board to study with T. H. Morgan.

There was a strategy behind Dobzhansky's "postdoc" in the Morgan lab: Dobzhansky was sent in order to become a quintessential Morganist so that he could return to head up *Drosophila* research in Leningrad. Thus, after Dobzhansky joined the Morgan lab, Filipchenko wrote several times urging him to remain there as long as possible in order to become a "splendid morganoid"—another Sturtevant, if not another Morgan.[37] Preparations for these future researches were begun on both sides of the ocean. The starting group was to include Iu. Ia. Kerkis, M. L. Bel'govskii, and N. N. Medvedev. From the outset, even with Dobzhansky abroad, Filipchenko asked him to direct all their research.[38] (Dobzhansky agreed, and in fact did so, from abroad, as long as possible.) For his part, Dobzhansky was planning to bring home many strains of *Drosophila*, and he expressed the hope that the collection would be the largest in Europe.[39]

---

[35] NRB, fund 813, delo 282, p. 104.

[36] "Otchet," 1926, pp. 225–226.

[37] NRB, fund 813, delo 1245, p. 107.

[38] NRB, fund 813, delo 1245, p. 148.

[39] NRB, fund 813, delo 283, p. 7.

In the meantime, Filipchenko was occupied in preparing the equipment and space for the *Drosophila* group. After Dobzhansky left for the States, Filipchenko was given some rooms for the Bureau of Genetics in a building on 6 Makarov Embankment (now the Physiological Institute of the Russian Academy of Sciences). At first the bureau occupied rooms with windows facing only the Makarov Embankment, but then rooms on the other side of the corridor were added. This is where Dobzhansky was slated to work. In short, Filipchenko made all the necessary preparations.[40] The intensive work was to begin immediately upon Dobzhansky's return.

CONCLUSION

Of course, Dobzhansky did not return. From December 1927, when Dobzhansky left for New York, until the spring of 1930, when Filipchenko died, things changed for both men. I have discussed elsewhere Dobzhansky's gradual decision to stay in the United States (Konashev 1991b).

As for Filipchenko, from 1925 through 1930 he found himself in the focus of the Communist Party's struggle for total control of Leningrad University. In 1925 Filipchenko and several other faculty members nominated their own candidate, K. B. Deriugin, for university rector and campaigned, unsuccessfully, for his election (V. B. Tomashevskii won instead). In 1926 Filipchenko headed an unsuccessful rebellion in the physics and mathematics faculty of the university against the appointment of the party candidate, K. A. Matsulevich, to the post of dean.[41] As a result of these election activities, Filipchenko was considered by the Party to be one of the leaders of the so-called "right professorate," concerning which the Leningrad Regional Party Committee (*gubkom*) sent a special message, signed by S. M. Kirov, to the Politburo (Izmozik 1991, p. 20).[42] Filipchenko died of meningitis in May 1930. It would not be too unfeeling to say that he died just in time.

If Dobzhansky had returned, it seems likely (as he always as-

---

[40] NRB, fund 813, delo. 1245, pp. 122–124. See also Gall and Konashev 1990.

[41] TsGA SPb, fund 7240, opis' 14, delo 178, pp. 118–181.

[42] See also M. B. Konashev. "The First University Department of Genetics in Russia (at Petersburg University): On the Right Road to Soviet Science, 1919–1930," in *Abstracts of the International Congress of the History of Science, Saragoza 1993* (in press).

sumed) that he would have shared the subsequent tragic fate of his friends G. D. Karpechenko and N. I. Vavilov. As we now know, Dobzhansky's student Kerkis was already under surveillance by the secret police (Kerkis 1989), and there is evidence that some other members of Filipchenko's group may also have been under surveillance.

Even before Dobzhansky became an American citizen, he was castigated in Russia as a *nevozvrashchenets* (nonreturner). In the 1940s he was referred to as an "enemy of the people," and then became simply a "nonperson." Many of his books and articles were removed from libraries and kept in the so-called *spetskhran* (restricted collections whose use required official clearance) (Konashev 1993a, 1993b). Only later, when it became possible, were Dobzhansky's relations with his friends and the scientific community in Russia restored (Gall and Konashev 1990, p. 87).

Even so, Dobzhansky's influence on the development of genetics in Russia was significant. During the mid and late 1920s Dobzhansky influenced biologists with his popular booklets and articles (Dobzhansky 1924d, 1925, 1926a, 1926b). His running dispute with Filipchenko over the relevance of genetics to evolutionary theory was well known, both inside and outside Filipchenko's group: in contrast to Filipchenko, Dobzhansky specifically emphasized the evolutionary meaning of mutations (Dobzhansky 1924b, p. 424). And although he did not have time to create his own school of genetics in Leningrad, he did have followers in the person of Kerkis, Bel'govskii, Medvedev, and Goroshchenko. Even during Stalinism, owing to his special status as a distinguished émigré, Dobzhansky continued to play a significant role in the development of genetics and evolutionary theory in the USSR.

ACKNOWLEDGMENTS

I wish to thank the archivists in Kiev, St. Petersburg, and Philadelphia who helped me find new documents concerning Dobzhansky. For their oral reminiscences of Dobzhansky, I am grateful to Ia. L. Goroshchenko, Iu. I. Polianskii, and S. M. Gershenzon. I wish especially to thank Sophia Dobzhansky Coe for making available a copy of the typewritten text of her father's reminiscences and Mark B. Adams, Nikolai L. Krementsov, and Daniel A. Alexandrov for information and helpful advice.

References

Manuscripts and Archival Collections

GAK—Gosudarstvennyi arkhiv goroda Kieva [Kiev city archive], fund 176, opis' 2, delo 24.

NRB—Natsional'naia Rossiiskaia Biblioteka, Rukopisnyi otdel [National Russian Library (formerly Saltykov-Shchedrin Metropolitan public library of St. Petersburg), department of manuscripts], fund 813, delo 282–284 and 1245.

PA RAN—Sankt-Peterburgskii Arkhiv Rossiiskoi Akademii Nauk [St. Petersburg archives of the Russian Academy of Sciences], fund 132, opis' 1, delo 45; fund 291, opis' 2, delo 52.

TsGA SPb—Tsentral'nyi gosudarstvennyi arkhiv goroda Sankt-Peterburg [Central state archives of the city of St. Petersburg], fund 7240, opis' 14, delo 142, 144, 152, 154, 178, and 193.

TsGANTD—Tsentral'nyi gosudarstvennyi arkhiv nauchno-tekhnicheskoi dokumentatsii goroda Sankt-Peterburg [Central state archive of scientific and technical documentation, St. Petersburg], fund 318, opis' 1–1, nos. 325, 368, and 371.

TsGB ANUk—Tsentral'naia gosudarstvennaia biblioteka Akademii nauk Ukrainy [Central scientific library of the Ukrainian Academy of Sciences, department of manuscripts], fund 175, delo 1429.

Books and Articles

Adams, Mark B. 1968. "The Founding of Population Genetics: Contributions of the Chetverikov School, 1924–1934." *Journal of the History of Biology* vol. 1, no. 1, pp. 23–39.

———. 1980a. "Sergei Chetverikov, the Kol'tsov Institute, and the Evolutionary Synthesis." In *The Evolutionary Synthesis: Perspectives on the Unification of Biology*, ed. E. Mayr and W. B. Provine (Cambridge: Harvard University Press), pp. 242–278.

———. 1980b. "Severtsov and Schmalhausen: Russian Morphology and the Evolutionary Synthesis." In *The Evolutionary Synthesis: Perspectives on the Unification of Biology*, ed. E. Mayr and W. B. Provine (Cambridge: Harvard University Press), pp. 193–225.

———. 1989. "The Politics of Human Heredity in the USSR, 1920–1940." *Genome* vol. 31, no. 2, pp. 879–884.

———. 1990a. "Eugenics in Russia, 1900–1940." In *The Wellborn Science: Eugenics in Germany, France, Brazil, and Russia*, ed. Mark B. Adams (New York: Oxford University Press), pp. 153–216.

———. 1990b. "Filipchenko, Iurii Aleksandrovich." In *Dictionary of Scientific*

Biography, vol. 17, supp. 2, ed. Frederic L. Holmes (New York: Charles Scribners' Sons), pp. 297–303.

———. 1990c. "Levitskii, Grigorii Andreevich." In *Dictionary of Scientific Biography,* vol. 18, supp. 2, ed. Frederic L. Holmes (New York: Charles Scribners' Sons), pp. 549–553.

Dobzhansky, Theodosius [Dobrzhanskii, F. G.]. 1922. "Otchet o rabotakh entomologicheskoi ekspeditsii po obsledovaniiu vreditelei svekly, snariazhennoi Kievskim oblastnym Upravleniem Sakharotresta vesnoi 1922 goda" [Report of the work of an entomological expedition investigating sugarbeet pests, supported by the Kiev Regional Department of the Sugar Trust, in spring 1922]. *Vist. Izukrov. Promyslennosti,* no. 5–6, pp. 11–115.

———. 1924a. "Über der Bau des Geschlechtsapparatus einiger Mutanten von *Drosophila melanogaster* Meig." *Zeitschrift für induktive Abstammungs- und Vererbungslehre* vol. 34, pp. 245–248.

———. 1924b. "Die geographische und individuelle Variabilität von Harmonia axyridis Pallas in ihren Wechselbeziehungen." *Biologische Zentralblatt* vol. 44, no. 8, pp. 401–426.

———. 1924c. "Die geographische und individuelle Variabilität von Adalia bipunctata L. und Adalia decempunctata L. (Coleoptera, Coccinellidae)." *Russkoe Entomologicheskoe Obozrenie* vol. 18, no. 4, pp. 201–212.

———. 1924d. "Nasledstvennost' i mutatsia" [Heredity and mutation]. *Chelovek i Priroda,* no. 5–6, pp. 417–426.

———. 1925. *Chto i kak nasleduetsia u zhivykh sushchestv* [What is inherited in living beings and how?]. Leningrad: Gosizdat.

———. 1926a. "Obzor geneticheskikh issledovanii vidov roda Drosophila" [Review of investigations on the genetics of species of the genus *Drosophila*]. *Trudy po prikladnoi botanike i selektsii,* vol. 15, no. 5, pp. 45–56.

———. 1926b. "K voprosu o nasledovanii priobretennykh priznakov" [On the problem of acquired characters]. In *Priformizm ili epigenezis?* [Preformism or epigenesis?], ed. E. S. Smirnov et al. Vologda: Severnyi pechatnik, pp. 27–47.

———. 1926c. "Dnevnik zaveduiushchego zhivotnovodcheskim otriadom Kazakhstanskoi ekspeditsii Akademii Nauk SSSR" [A diary of the animal breeding group of the Kazakhstan expedition of the USSR Academy of Sciences]. Manuscript.

———. 1927a. "Loshad' kochevogo naseleniia Semirech'ia" [Horses of the nomadic population of Semirechie]. In *Materialy Osobogo Komiteta po Issledovaniiu v Soiuznykh i Avtonomnykh Respublikakh, Akademiia Nauk SSSR* [Materials of the special committee of the USSR Academy of Sciences on research in union and autonomous republics], no. 8, pp. 16–131.

———. 1927b. "Studies on the Manifold Effect of Certain Genes in *Drosophila melanogaster."* *Zeitschrift für induktive Abstammungs- und Verer-bungslehre* vol. 43, pp. 330–388.

Dobzhansky, Theodosius. 1928. "Loshad' kochevogo naseleniia Semipala-tinskoi gubernii" [Horses of the nomadic population of the Semipala-tinsk province]. In *Materialy Osobogo Komiteta po Issledovaniiu v Soiuz-nykh i Avtonomnykh Respublikakh, Akademiia Nauk SSSR* [Materials of the special committee of the USSR Academy of Sciences on research in union and autonomous republics], no. 18, pp. 22–183.

————. 1962–1963. "The Reminiscences of Theodosius Dobzhansky." Typed transcript. 2 parts. Oral History Research Office, Columbia University, New York.

————. 1980. "The Birth of the Genetic Theory of Evolution in the Soviet Union in the 1920s." In *The Evolutionary Synthesis: Perspectives on the Unification of Biology,* ed. E. Mayr and W. B. Provine (Cambridge: Harvard University Press), pp. 229–242.

Ehrman, L., and B. Wallace. 1976. "Obituary." *Nature* vol. 260, p. 179.

Filipchenko, Iu. A. 1919. "Khromozomy i nasledstvennost'" [Chromosomes and heredity]. *Priroda,* no. 7–9, pp. 327–350.

————. 1922. "Zakon Mendelia i zakon Morgana" [Mendel's law and Mor-gan's law]. *Priroda,* no. 10–12, pp. 51–66.

Gall, Ia. M., and M. B. Konashev. 1990. "Klassik" [Classic]. *Priroda,* no. 3, pp. 79–87.

Inge-Vechtomov, B. G., ed. 1993. *Issledovaniia po genetike* [Researches in ge-netics], no. 11. St. Petersburg (in press).

Izmozik, V. S. 1991. "Problema sekretnosti v otnosheniiakh partiinogo ap-parata i nauchno-pedagogicheskoi intelligentsii v 20-e gody" [The prob-lem of secrecy in the relations between the party apparatus and the ped-agogical intelligentsia in the 1920s]. In *Svoboda nauchnoi informatsii i gosudarstvennaia sekretnost'* [Freedom of scientific information and state secrets] (Leningrad), pp. 18–20.

Kerkis, Iu. Ia. 1989. "Neizvestnye stranitsy iz zhizni N. I. Vavilova" [Unknown pages of N. I. Vavilov's life]. *Priroda,* no. 3, pp. 97–102.

Konashev, M. B. 1991a. "Dobrzhanskii: genetik, evoliutsionnist, gumanist" [Th. G. Dobrzhanskii: geneticist, evolutionist, humanitarian]. *Voprosy istorii estestvoznaniia i tekhniki,* no. 1, pp. 56–71.

————. 1991b. "Ob odnoi nauchnoi komandirovke, okazavsheisia bessroch-noi" [The postdoc that never ended]. In *Repressirovannaia nauka* [Sci-ence under repression], ed. M. G. Iaroshevskii (Leningrad: Nauka), pp. 240–263.

————. 1993a. "Kritika Dobrzhanskim lysenkoizma" [Dobrzhansky's criti-cism of Lysenkoism]. *Voprosy istorii estestvoznaniia i tekhniki* (in press).

————. 1993b. "F. G. Dobrzhanskii i stanovlenie genetiki v Leningradskom Universitete" [Th. Dobzhansky and the origin of genetics in Leningrad

University]. In *Issledovaniia po genetike* [Researches in genetics], no. 11. St. Petersburg (in press).

——. 1993c. "Lysenkoizm pod okhranoi spetskhrana" [Lysenkoism under the protection of *spetskhran*]. In *Repressirovannaia nauka* [Science under repression], vol. 2 (in press).

Medvedev, N. N. 1978. *Iurii Aleksandrovich Filipchenko.* Moscow: Nauka.

Morgan, T. H., A. H. Sturtevant, H. J. Muller, and C. B. Bridges. 1915. *The Mechanism of Mendelian Heredity.* New York: Henry Holt & Co.

Nauchnye uchrezhdeniia Leningrada [Scientific institutions of Leningrad]. 1926. Leningrad.

Naumov, G. 1989. "F. G. Dobrzhanskii (1900–1975) i sovetskaia genetika (svetloi pamiati velikogo biologa)" [Th. Dobzhansky and Soviet genetics: in bright memory of a great biologist]. *Genetika*, no. 6, pp. 1131–1135.

Otchet o deiatel'nosti Akademii nauk SSSR za 1925 g. [Report on the activity of the USSR Academy of Sciences for 1925]. Leningrad: 1926.

Prokof'eva-Bel'govskaia, A. A. 1978. "Predislovie" [Preface]. In *M. E. Lobashev i problemy sovremennoi genetiki* [M. E. Lobashev and problems of modern genetics] (Leningrad: Leningrad University Press), pp. 3–6.

Raipulis, E. P. 1985. "Zhizn' i deiatel'nost' Ia. Ia. Lusisa" [The life and work of J. J. Lusis]. In *Ia. Ia. Lusis: zhizn' i nauchnaia deiatel'nost'* [Ia. Ia. Lusis: life and scientific activity] (Riga: Zinatine), pp. 13–87.

Sorokina, M. Iu. 1990. "Dal'nii put' k bol'shomu budushchemu: Iz perepiski F. G. Dobrzhanskogo s. V. I. Vernadskim" [The long path into the big future: from the correspondence of F. G. Dobrzhanskii and V. I. Vernadskii]. *Priroda*, no. 3, pp. 88–96.

Vavilov, N. I. 1990. *Zhizn' korotka, nado speshit'* (Life is brief, we must hurry). Moscow: Sovetskaia Rossia.

PART TWO
THE MORGAN LAB

Theodosius Dobzhansky, the Morgan Lab, and the Breakdown of the Naturalist/Experimentalist Dichotomy, 1927–1947

Garland E. Allen

WE ARE ALL familiar with Theodosius Dobzhansky's statement "Nothing in biology makes sense except in the light of evolution" (Dobzhansky 1972). I would like to slightly amend that quotation with what I hope would be Professor Dobzhansky's approval: "Nothing in biology makes sense except in the light of evolution *and genetics.*" I suggest this modification because it emphasizes the two most crucial aspects of Dobzhansky's work. One is the application of new genetic techniques, during the period of the "evolutionary synthesis," to the solution of problems in evolutionary theory. More fundamentally, the modification also gives equal recognition to the two previously separate traditions that Dobzhansky brought together most successfully in his work: laboratory, experimental work on the one hand and field natural history on the other. It is, indeed, the specific fusion of Mendelian, particulate genetics with the Darwinian theory of natural selection, and the more general fusion of the laboratory and field naturalist traditions, that remains among the deepest and most lasting aspects of Dobzhansky's legacy.

In this paper I would like to examine what role Dobzhansky's experience in the laboratory of Thomas Hunt Morgan, both at Columbia University (1927–1928) and later at Caltech (1928–1940), played in Dobzhansky's bringing together the methods of the field naturalist and laboratory experimentalist. Much of what I have to say comes from my study of the history of genetics and of evolutionary theory in the twentieth century. But much of it is also drawn from a formal interview with Dobzhansky in October 1966 and an informal discussion with him two years later in the fall of 1968. I also have drawn heavily on the work of John Beatty (Beatty 1987) and of Richard Lewontin, William Provine, John Moore, and Bruce Wallace, who provided an immense service to the biological and historical communities by reprinting the "Genetics of Natural Population" series in 1981 (Lewontin et al. 1981). Lewontin's "Introduction" and Provine's lengthy historical analysis are valuable source materials.

In the mid-1960s I was anxious to meet Dobzhansky for several reasons. First, I had admired him ever since taking a seminar in evolutionary biology with Ernst Mayr in 1961, where Dobzhansky's work received considerable attention. Second, Dobzhansky had come to the United States in December 1927 expressly to work in the laboratory of Thomas Hunt Morgan at Columbia University, and in 1966 I was beginning to work seriously on a biography of Morgan (Allen 1978). I was traveling about trying to meet all the people who had studied under or worked with Morgan, and that naturally led me to Dobzhansky's doorstep.

Dobzhansky was a particularly likely candidate to interview. First, he was interested in evolutionary theory, and Morgan had always had a rather eclectic approach to evolution, at least early in his career (prior to 1915 or 1916), being highly skeptical of Darwin's theory of natural selection. Although Morgan had more or less accepted the Darwinian theory by the time Dobzhansky arrived in New York, he still had some reservations, and Muller, among others, claimed that Morgan never really understood the basic tenets of natural selection. Thus, it seemed that Dobzhansky might have had more discussions with Morgan on evolutionary topics than had appeared to be the case with other members of the lab, such as Sturtevant or Muller, whom I had already interviewed. As my interview showed, Dobzhansky had indeed turned out to be particularly close to Morgan in those years and provided a much different perspective from Sturtevant, the "insider" and Morgan student. Dobzhansky told me, for example, that he and Morgan used to discuss religion a lot because, as he put it,

> Morgan's idea about what science was for was to dispel mystery. Mystery is something which feeds religion. The point of science is to deprive religion of this source of support. That is what all science is for. And that is, if you please, what evolution and genetics is for. . . . Evolution and genetics dispels the mystery of the origin of the world and the origin of man. . . . By 1931 or 1935 . . . Morgan felt that in a sense evolution and genetics had done their job. Heredity and evolution are no longer mysteries. . . . (Allen 1967: 26)

Dobzhansky claimed that Morgan liked to discuss religious and quasi-philosophical issues with him in particular because the others in the lab, namely Bridges and Sturtevant, tended to avoid such topics. They shared Morgan's views about religion but apparently did not enjoy such discussions. Dobzhansky obviously did, and for this reason, among many others, from Morgan's perspective Dobzhansky was a specially welcome addition to the group. He was certainly a welcome addition to my pool of interviewees because he seemed to bring out a side of Morgan that others in the lab did not: an intense interest in how genetics related to the larger field of biology—especially evolution and development—and in philosophical matters. As Dobzhansky pointed out to me, Morgan's strong dislike of religion was only a part of his more general distaste for metaphysics—the abstract, speculative, and nondemonstrable. The "Morganization" of Dobzhansky, about which I will speak in a moment, was greatly facilitated by the fact that both men shared an interest in the wider world of biology, and even philosophy, that was lacking in the other immediate members of the Morgan group in the period after 1928.

BACKGROUND TO THE "MORGANIZATION" OF THEODOSIUS DOBZHANSKY

I will examine first the question of how Dobzhansky managed to fuse the field natural history tradition (one within which Darwin and his successors worked so well) with the laboratory tradition exemplified by the Morgan group. To understand the significance of this change, we need to step back a moment and look at the dichotomy that existed in Western science between the laboratory and field traditions. The depth of that distinction, and the social and institutional forms it took, had produced a major division between two large groups within the biological community: the experimentalists and the naturalists.

I have described what I see to be the extent of this division in a number of earlier papers, so I will not go into a lengthy discussion here (Allen 1979). The naturalist tradition is that of Linnaeus, Buffon, Lamarck, and Darwin. Naturalists were museum- and field-oriented biologists who worked with descriptive methods and were concerned with the life habits, taxonomic patterns, and historical origins of organisms. They had a feel for animals and plants in their natural habitats—for what we today would call their ecological relationships. Naturalists tended to base their conclusions on observational evidence, looking at nature more "as a whole" than at its component parts in isolation. Although naturalists were often highly precise and attentive to detail, they were nonanalytical and not rigorous in the usual sense of those terms—that is, in modern, or Popperian language, they could not easily falsify their hypotheses. Thus, their conclusions—about taxonomic groupings or phylogenetic relationships—were largely inferential and speculative. For example, a good case could be made for either the annelid or arthropod origins of the Pycnogonids (sea-spiders), with no firm way to distinguish one alternative from another.

The experimentalist tradition, by contrast, presented a very different outlook. Experimentation, as a methodology, gained its great impetus in the seventeenth century and was first and foremost concerned with manipulating natural systems to answer specific questions. Experimentalists are interventionists, the experimenter deliberately altering the natural condition to observe whatever the consequence might be. Common to all types of experimentation has always been an emphasis on precision, measurement (quantitative as opposed to qualitative data), and analytical thinking (the isolation of parts of a complex whole and investigation of one at a time). A common assumption behind experimentation is the conviction that intervention into the workings of the organism, although artificial, still reveals something about natural or normal states of nature. The experimental tradition has been more closely associated with the investigation of proximal or functional, rather than with ultimate or historical, questions. Among the life sciences, physiology and biochemistry are preeminent examples of the experimentalist tradition.

The naturalist and experimentalist traditions have been—and in some respects still are—often viewed as mutually exclusive and antagonistic methodologies or philosophies of research. Naturalists accuse experimentalists of dealing with artificial circumstances, with conditions so abnormal that they throw no light on the natural process. Naturalists also claim that experimentalists have no appreciation of the large picture, of animals and plants as whole organisms or as parts of natural populations and communities. They feel that experimentalists do not understand the organism as a whole. In addition, naturalists distrust, or are sometimes afraid of, quantitative and especially mathematical methods and thus find the work of experimentalists abstruse and unapproachable.

For their part, experimentalists have often thought that naturalists are fuzzy and subjective thinkers, substituting a "feeling for the organism" for hard-nosed, rigorous investigation. In the past they have decried naturalists' reliance almost solely on descriptive methodology as leading only to speculative conclusions. Moreover, sometimes identifying the naturalist tradition with taxonomy, experimentalists have often claimed that naturalists are nothing more than stamp collectors.

The distinctions between the naturalist and experimentalist traditions have been social and institutional as well as intellectual. In the early parts of the century, especially in the United States, university as well as museum positions in biology were most frequently held by naturalists, whereas the relatively few experimentalists were often housed in medical schools or served as the one physiologist within a zoology department. However, by the first decade of the century experimentalists began gaining power and prestige within university zoology and botany departments, a process greatly accelerated by the rise of experimental embryology (or experimental morphology, as it was sometimes called) and by Mendelian genetics after 1910 (Allen 1985). New funding sources, such as the Rockefeller and Carnegie foundations and their subsidiaries, began support of the new genetics and experimental breeding work on a larger scale than naturalists of an earlier generation had usually enjoyed. In competition for funds, students, and prestige, the two traditions clashed repeatedly, though it is fair to say that the offensive was usually taken by the experimentalists. The conflict still lingers on today, surfacing in the heated debates between cladists and evolutionary taxonomists and between molecular biologists and ecologists. Happily, the distinctions and rivalry are beginning to break down as evolutionary biologists and taxonomists use molecular methods to great advantage and as molecular biologists become interested in historical questions such as evolution and the origin or life. But the rift has been a long-standing and deep one, producing a gap between two important and, from our perspective today, complementary methodologies within the biological community. Bridging the gap between field and laboratory traditions was begun by numerous workers earlier in the century, but no one did so more effectively than Theodosius Dobzhansky.

Dobzhansky's Background and the Morgan Lab

As Dobzhansky explained to me in our interview in 1966, he came from a clearly natural history background which, as Daniel Todes (Todes 1989) has shown, was highly imbued in Russia at the time with a Darwinian perspective:

> You see, in Russia, evolution was always considered to be a matter of utmost importance, both philosophically and even sociologically. You may or may not know but to a number of Russian naturalists and thinkers in the nineteenth century Darwin was not just a scientific theory but the basis for a whole philosophy. (Allen 1967: 3)

Dobzhansky was fortunate in that his early training, though largely in natural history, was fused with both Darwinian theory and with a respect for "experimental zoology." His teacher at Kiev, the cytologist Kushakevich, called himself an "experimental zoologist," and though he did not discuss formal genetics much, Kushakevich appeared to be sympathetic to such work as was to emerge from genetics laboratories by the end of the war (unfortunately he died in 1920 and never knew about the new work emerging from laboratories such as Morgan's, Bateson's, and others'). Kushakevich was, however, sympathetic to experimental work—after all, he had been a student of Richard Hertwig in Munich, from whom he gained an interest first in the problem of sex determination, working on frogs, and later on snails. Thus, from his earliest training Dobzhansky was prepared to see the importance of functional and experimental questions as they related to the evolutionary process.

The framework of Dobzhansky's own early work on polymorphism in coccinellid beetles was clearly evolutionary: trying to determine the amount of variation within natural populations, a question that bore directly on the rate and extent at which selection could operate (Lewontin et al. 1981: 95). One of the major objections to natural selection as a mechanism for the origin of species had always been that variations occurred too infrequently, and were too minute, to serve as the raw material for species-level changes. From his early work, Dobzhansky concluded that species were not largely homogeneous "types" but were, as Lewontin describes it, "Mendelian populations—that is, a group of individuals containing large amounts of genetic variation and exchanging genes ..." (Lewontin et al. 1981: 96). The problem was, of course, in 1926 or 1927, to estimate accurately the amount of genetic variation in a population in nature. Especially because of World War I and the disruptions caused by the Bolshevik Revolution, foreign scientific literature was scarce in the Soviet Union and Dobzhansky was slow to learn about the advances in genetics—particularly the introduction of *Drosophila* and the study of mutations—going on in England, Germany, Scandinavia, and the United States.

According to Dobzhansky, it was Vavilov who brought back large quantities of journals and other materials from the West after the war and who exposed Russian biologists to the new genetics. The materials were available in only two places, Dobzhansky reported: at Koltsov's institute in Moscow and at Filipchenko's institute in Leningrad. It was at the latter that Dobzhansky began to learn about the work of the Morgan group, largely, he reported, through a review article written by Filipchenko in the Russian journal *Priroda* in the early 1920s. As Dobzhansky said, "That was a sort of revelation," and was one of the reasons he went to Leningrad as Filipchenko's assistant in 1924. Dobzhansky described his arrival and experience in Leningrad succinctly: "I arrived [in Leningrad] on the night of Lenin's death, so I was probably three and one-half years with Filipchenko there. And of course, by that time, genetics was *it*. Morgan was a hero or a saint" (Allen 1967: 5).

When the opportunity came to take up a Rockefeller Foundation Fellowship in 1927, Dobzhansky was convinced he had to go to New York to learn about the mechanism of genetic transmission. In this he saw a possible way to investigate at a more precise and discrete level of organization the problem of variation in natural populations. What Dobzhansky brought with him to New York was thus a thorough grounding in evolutionary theory combined with extensive experience as a field naturalist. At the same time, because of the Russian milieu, Dobzhansky saw no inherent conflict between field studies—for example, of phenotypic polymorphism—and laboratory studies of the architecture of the germ plasm—that is, experimental, quantitative, and analytical work in Mendelian genetics.

In New York Dobzhansky soon found that he had landed in the midst of a very exciting laboratory. Accustomed to the European style of younger investigators working on a problem laid out by the head of the laboratory, Dobzhansky immediately approached Morgan on his first or second day at the laboratory and asked what the lab was working on and what he should investigate. As Dobzhansky describes it, the result was not what he expected:

Coming from Europe, coming to Morgan himself, I certainly thought I must get to the newest, most active work and asked what he would like me to do ... he was sort of indefinite. So in my very poor English I repeated that [question] several times. After several repetitions he pushed a box toward me and he said "This." So, I went home and my so-to-speak hair stood on end. It was perfectly respectable work [by Chester Bliss, later at Yale University, whose work in the late 1920s dealt with mathematical analyses of rates of development in *Drosophila*, physical chemistry, and the like], but it had nothing to do with genetics. It didn't interest me in the least. My wife and I still remember the most uncomfortable night which we passed thinking, well, what do we do now? Well, of course, very quickly I saw that ... he didn't in the least intend to urge me to do just that [kind of work]. In fact, he clearly gave the thing to me to get rid of [me].... (Allen 1967: 20)

Soon Dobzhansky had embarked on several experiments studying the effects of temperature on sex ratios and the like, realizing that Morgan insisted that people work on problems that interested *them*. This is why Dobzhansky was free to pursue his interests in genetics as it grew out of his primary interest in evolution. The relation between these was clear in Dobzhansky's mind at the time he entered the Morgan group:

My interest in genetics came from my interest in evolution—[my] interest in evolution, I may say, was philosophical, it came first; interest in genetics came from it. So, I had no doubt from the start that this is what I wanted to get into. (Allen 1967: 21)

And in Morgan's lab at Columbia, but especially at Caltech, Dobzhansky was able to pursue his own interests for two reasons: (1) Morgan encouraged that; and (2) Bridges and Sturtevant introduced him to the techniques of chromosome analysis. The latter was to provide him with a far more precise and rigorous method of detecting variability—particularly using chromosome morphology—than was possible by inspecting only gross phenotypes. Phenotypic-level traits often conceal rather than reveal variation, a problem that was an immense barrier to understanding evolution in field populations. It was a problem that Dobzhansky was determined to try and resolve.

But first he had to resolve two other problems he had not expected to encounter. The first was that there was a prevailing attitude within the Morgan group that experimental work was superior to nonexperimental work, that is, to purely descriptive and/or field studies (Allen

1978: 318ff). Even though—and perhaps, as I have argued elsewhere, because—Morgan himself had been trained as a descriptive morphologist, he and other members of the group tended to look down on descriptive or field work. *Natural history* was a bad term in the Morgan group. They tended to see it as "soft" science: qualitative, descriptive, nonrigorous, and speculative. The catchword was that such work was "stamp collecting" and could not explain any of the important problems in biology. Of the main Morgan group, Sturtevant had the best appreciation of field studies and was the most accomplished *Drosophila* taxonomist among them. Despite this, however, field studies were seen to be the opposite of laboratory work, and decidedly inferior. To Dobzhansky, for whom no such dichotomy existed, this attitude was a stumbling block.

Associated with its opposition to field and descriptive work, the Morgan group also harbored a strong skepticism about current studies in evolution. Although by 1928 Morgan had overcome his earlier skepticism about Darwinian theory, a skepticism the younger members of the group never had, there was still a prevalent feeling that all evolutionary studies were mired in speculation and guesswork and could never hope to provide any rigorous answers to the problem of how evolution really came about. Indeed, to the Morgan group, the prototype of current evolutionary theory was embodied in the work of Henry Fairfield Osborn (1857–1935), president of the American Museum of Natural History in New York and professor of paleontology at Columbia. With his theory of "aristogenesis" and other hypothetical forces moving evolutionary development in certain directions, Osborn appeared to Morgan and his group to be the epitome of the futile nature of such an approach to biology (Allen 1978: 320–321). Osborn and other evolutionary theorists were thought to be old-fashioned, armchair speculators, field naturalists thoroughly lacking in the modern spirit of quantitative and rigorous biology.

Thus, Dobzhansky found himself up against a double barrier—anti–field methods and anti–evolutionary theory—within the Morgan group. The prevailing attitude was particularly distressing to Dobzhansky, for whom there was no necessary dichotomy between field and laboratory studies on the one hand, or between evolutionary theory and the rest of biology on the other. Nevertheless, Dobzhansky was able to take advantage of what the Morgan group had most to offer: experimental techniques and methods of research with *Drosophila*. As has been discussed by others in the present volume, one of the most important techniques Dobzhansky learned from the Morgan group was that of chromosome analysis. Especially after the discovery of the giant salivary chromosome of *Drosophila* in the early 1930s, Dobzhansky found that he could use the band patterns and other visual markers as a precise indicator of genetic variability within and between populations. It was this technique, and its development by Dobzhansky over the years, that formed the basis of the "Genetics of Natural Populations" series.

Despite the prevailing attitude within the Morgan group, Dobzhansky was not quite alone in pursuing his interest in evolutionary problems or in carrying out field studies. As William Provine has shown (Provine 1981), it was especially the collaboration with Sturtevant, after the group moved to Caltech, that helped get Dobzhansky going on his research program. Provine has also shown that Dobzhansky slowly learned, from his interactions with Sewall Wright (1889–1988), that field experiments could be organized with the same precision and rigor that was characteristic of laboratory experiments. This was a major step in Dobzhansky's later demonstration that field and experimental work could be effectively combined.

Once the older distinction between field natural history and experimental laboratory work began to be eroded, other investigators followed Dobzhansky's lead. In the 1950s, for example, Hampton Carson, in collaboration with Harrison Stalker, began to apply Dobzhansky's chromosome polymorphism methods to investigate the problem of island colonization in the Hawaiian *Drosophila* (Carson 1971, 1980, 1982). And in the mid-1960s Richard Lewontin and J. L. Hubby introduced gel electrophoresis as a new method of detecting genetic polymorphism in proteins both within and between populations (Hubby and Lewontin 1966). These approaches, as just two of many examples, indicate the extent and significance of Dobzhansky's influence on the unification between genetics and evolutionary theory on the one hand and between the naturalist and experimentalist traditions on the other.

CONCLUSIONS

What I have suggested in this paper is that while Dobzhansky took from his experience with the Morgan group an extremely important technique—the analysis of chromosome fine structure—he contributed something of equal or greater importance. That contribution was the conviction that the experimental and naturalist traditions were in no way contradictory or mutually exclusive. Indeed, in Dobzhansky's hands they complemented and reinforced each other. The specific cytogenetic techniques concerned with chromosome analysis and a recognition of the importance of using chromosome structure as a means of analyzing the amount of variation in populations that Dobzhansky learned from the Morgan group were invaluable to his further work in evolutionary theory. What was so fortunate about *Drosophila* as a favorable organism, for Dobzhansky's purposes, was that it was one of the organisms that could be collected easily in the field and yet whose genetics was well understood in the laboratory. Prior to his work with the Morgan group, Dobzhansky had been frustrated by his field studies of the Coccinellidae because so little was known, or could be determined easily, about their genetics. This was not a problem with *Drosophila.*

Ultimately, however, Dobzhansky contributed to Western biology something far more important than the specific techniques that he developed to analyze chromosome polymorphism, or even the application of these techniques to natural populations. That contribution was the beginning of the breakdown of the long-standing dichotomy—indeed, antagonism—between field and laboratory work, between the naturalist and the experimentalist traditions. His ability to accomplish this task was a product of his uniquely Russian background, in which the naturalist-experimentalist dichotomy did not seem to exist, at least in his own teachers or in anything like the degree it existed in the West (Adams 1970, 1980). The view that field and laboratory work were compatible, even complementary, and that one must combine both in order to study the evolutionary process, has been one of the most important aspects of Dobzhansky's legacy to modern evolutionary theory and indeed to all of modern biology.

REFERENCES

Adams, Mark B. 1970. "Towards a Synthesis: Population Concepts in Russian Evolutionary Thought, 1925–1935." *Journal of the History of Biology* 3: 107–129.
———. 1980. "Science, Ideology and Structure: The Kol'tsov Institute, 1900–1970." In Linda L. Lubrano and Susan Gross Solomon, eds., *The Social Context of Soviet Science* (Boulder, Colorado: Westview Press), pp. 173–204.

Allen, Garland E. 1967. "Interview with Theodosius Dobzhansky, October, 1967." Typescript transcribed and edited by G. E. Allen.

————. 1978. *Thomas Hunt Morgan: The Man and His Science*. Princeton, N.J.: Princeton University Press.

————. 1979. "Naturalists and Experimentalists: The Genotype and the Phenotype." In W. Coleman and C. Limoges, eds., *Studies in History of Biology*, vol. 3 (Baltimore: The Johns Hopkins University Press), pp. 179–209.

————. 1985. "T. H. Morgan and the Split Between Embryology and Genetics, 1910–1935." In T. Horder and J. A. Witkowski, eds., *A History of Embryology* (New York: Cambridge University Press), pp. 113–146.

Ayala, Francisco. 1977. "'Nothing in biology makes sense except in the light of evolution'." *Journal of Heredity* 68: 3–10.

Beatty, John. 1987. "Dobzhansky and Drift: Facts, Values, and Chance in Evolutionary Biology." In Lorenz Kruger, Gerd Gigerenzer, and Mary Morgan, eds., *The Probabilistic Revolution*. Vol. 2, *Ideas in the Sciences*. Cambridge: MIT Press.

Carson, Hampton L. 1971. "Speciation and the Founder Principle." *Stadler Genetics Symposium* 3: 51–70.

————. 1980. "Cytogenetics and the NeoDarwinian Synthesis." In Ernst Mayr and William Provine, eds., *The Evolutionary Synthesis: Perspectives on the Unification of Biology* (Cambridge: Harvard University Press), pp. 86–95.

————. 1982. "Hawaii: Showcase of Evolution. An Introduction." *Natural History* 91: 16–18.

Dobzhansky, Theodosius. 1962–1963. "The Reminiscences of Theodosius Dobzhansky." Typed transcript. 2 parts. Oral History Research Office, Columbia University, New York.

————. 1972. "Nothing in Biology Makes Sense Except in the Light of Evolution." *American Biology Teacher* 35: 125–129.

Hubby, J. L., and R. C. Lewontin. 1966. "A Molecular Approach to the Study of Genic Heterozygosity in Natural Populations." *Genetics* 54: 577–594.

Lewontin, R. C., J. A. Moore, W. B. Provine, and B. Wallace, eds. 1981. *Dobzhansky's Genetics of Natural Populations I–XLIII*. New York: Columbia University Press.

Provine, William B. 1981. "Origins of the 'Genetics of Natural Populations' Series." In R. C. Lewontin, J. A. Moore, W. B. Provine, and B. Wallace, eds. *Dobzhansky's Genetics of Natural Populations I–XLIII* (New York: Columbia University Press), pp. 5–83.

————. 1986. *Sewall Wright and Evolutionary Biology*. Chicago: University of Chicago Press.

Todes, Daniel P. 1989. *Darwin Without Malthus: The Struggle for Existence in Russian Evolutionary Thought*. New York: Oxford University Press.

■

The Origin of Dobzhansky's
Genetics and the Origin of Species

William B. Provine

INTRODUCTION

The year 1936 was momentous for Theodosius Dobzhansky. He began the year as an assistant professor of biology at the California Institute of Technology (Caltech) in Pasadena, collaborating with his senior colleague A. H. Sturtevant on the most exciting work he had ever done in science—investigating chromosomal inversions in natural populations of *Drosophila pseudoobscura* and its close relatives. Early in the year an attractive offer arrived from Milislav Demerec, head of the *Drosophila* research program in the Department of Genetics at the Station for Experimental Evolution in Cold Spring Harbor, for Dobzhansky to spend several months of the summer doing his research in Cold Spring Harbor.

Dobzhansky then received an invitation from Columbia University geneticist Leslie C. Dunn to visit Columbia for six weeks to inaugurate a new set of Jesup Lectures, a series dormant since 1910. Soon thereafter, geneticist J. T. Patterson wrote from the University of Texas to offer Dobzhansky the full professorship vacated by H. J. Muller, at a salary twice that Dobzhansky was receiving as an assistant professor at Caltech. After Dobzhansky wrote Demerec about the Texas offer, Demerec admitted that his real hope was to hire Dobzhansky as a full-time researcher at Cold Spring Harbor. By the middle of the year, Thomas Hunt Morgan engineered Dobzhansky's promotion to the rank of full professor at Caltech with his salary nearly doubled.

In October and early November Dobzhansky gave the Jesup Lectures, and they were very well received. In late December the newly reconstituted board of editors for the Columbia Biological Series at Columbia University Press decided in principle to publish the lectures as a book. In the last few days of 1936 Dobzhansky began writing, and he finished the book within months. It was quickly published, and it made him one of the most prestigious evolutionary biologists in the world. A momentous year indeed. (For a more detailed background on this year, see the rest of this volume and also Provine 1981.)

This essay will trace the origin of Dobzhansky's *Genetics and the Origin of Species* in the context of the tumultuous year described above. I have drawn upon a wealth of source materials, including the Columbia University oral memoirs of both Dobzhansky and Dunn and a great deal of contemporary correspondence between the major participants, as well as recollections by Ernst Mayr, John A. Moore, and Kenneth Cooper, all of whom were in New York City at the time and attended the Jesup Lectures. Nevertheless I will raise many questions for which I have no certain answers and sometimes even little circumstantial evidence.

DUNN'S RECOLLECTION

In his oral memoir, Dunn described his early impressions of Dobzhansky, the Jesup Lectures, and *Genetics and the Origin of Species.*

> I became first acquainted with him in 1936, when we decided here to revive the annual lectureship known as the Jesup Lectures. This had been started originally by H. F. Osborn and it was named in honor of Morris K. Jesup, who had been a benefactor of our Museum of Natural History, a banker, and in fact, when Osborn got the thing started I'm sure he anticipated that Mr. Morris K. Jesup was going to remember it in his will because he had a lectureship named for him. Unfortunately, Jesup forgot that.
>
> So the lectures had run and somehow lapsed, we don't know quite why, from about 1910, but looking back over some of the old books that were published as a part of that series, we decided it was too good to let go, and that problems of evolution had taken such a new and different form, due to the influence of modern genetics, that we would re-inaugurate the series, which in general had dealt with questions of evolution, by facing the problem most directly. So, we invited a set of lectures from Dobzhansky with a very bold title: Genetics and the Origin of Species—to indicate the main topics that were to be dealt with. The title, in fact, we thought might be a little too ambitious, but Morgan, when we spoke to him about our purpose in inviting Dobzhansky, said, "A bold title would best express what this fellow will talk about."
>
> So it was retained, and he came and gave those lectures, and was in residence here during the spring term of 1936 [actually October and

early November]. Then the book was prepared for publication and appeared under the title, Genetics and the Origin of Species. Dobzhansky was the one who carried most eagerly, I should say, the program of testing Darwin's ideas by actual experiments. He thought, if natural selection is a fact, you should be able to detect changes in populations of rapidly reproducing animals. Moreover, you should be able to follow these changes in such populations in nature. And it's because he had been doing this that we thought of him for our lectureship.

The lectures and the book were such a clear indication of what the future might be for that kind of study that after two years we invited him to come as professor and he accepted and began his professorship here in 1940, and inaugurated the period of association which has been unbroken up to the present.

Up until this time, although I'd had let's say an ancillary interest in the question of the application of genetical ideas to evolutionary problems, my interests had never taken very precise form. It was Dobzhansky's influence, I think, which turned my interests in a more operative way towards that field. I should say that my generation of graduate students was warned by the professors not to get too involved in questions of phylogeny and of evolution, because these were likely to end up in speculation, and their opinion—and it was a fairly general one among experimental biologists—was that speculation might be a brake on these. Of course, it's turned out in the opposite way; certain speculative ideas have to be thought of before one knows what experiments to apply to them. . . .

At any rate, Dobzhansky certainly stimulated my interest in evolution and in population genetics. At that time, population genetics was one of those fields in which theory had far outrun facts. . . . [Explains the work of Fisher, Haldane, and Wright.]

So the theoretical structure was available. That was too mathematical for most people, but it had been to some extent made understandable and palatable, especially by Fisher, and Dobzhansky's contribution, I think, was to take the mathematical theory and test it by observations on populations in nature. That's what his book was about. It's the evidence for the main factors in evolution: mutation, natural selection, sampling in small populations (which Wright had called random genetic drift) and forces which tend to change populations with reference to each other, such as migrations. (Dunn 1961, pp. 865–869)

DOBZHANSKY'S RECOLLECTION

In a section describing the events of 1936, Dobzhansky first detailed his invitation from his "lifelong" friend Demerec to go to Cold Spring

Harbor for several months in the summer of 1936. Then he turned to the Jesup Lectures and *Genetics and the Origin of Species.*

Simultaneously, or almost simultaneously, my other lifelong friend, L. C. Dunn, of the Department of Zoology, Columbia University, invited me to give the so-called Jesup Lectures in the Department, to be written up later in the form of a book published by Columbia University Press. I have chosen as a topic for the lectures, "Genetics and the Origin of Species." The lectures were given in October and early November, 1936. . . .

In the end of September or maybe early October, [we] moved to New York, took an apartment on 110th Street, near Columbus Avenue, which at that time was not a pleasant place to live. The elevated line was still on. Our windows I think on the 3rd or 4th floor were overlooking 110th Street and the elevated tracks. Besides, the apartment was pretty dirty and unpleasant. We lived there however for, I think, it was a month or six weeks, during which I was giving the Jesup Lectures in the Department of Zoology, Columbia University. These were the first Jesup Lectures after a long interval.

Now, the Jesup Lectures are sometimes given twice a year, but usually once a year. The stipend, the payment, is pretty skimpy. At that time it was barely enough to pay for our life in New York. At present it pays something like $600. I believe I got less. . . .

The Jesup Lectures, according to the conditions of the lectures, requires a book to be written, presented, and published by Columbia University Press which, since 1936, is a condition which has been fulfilled, shocking to say, by approximately one out of every three lecturers. [Gives example of Silliman Lectures at Yale where stipends are not paid until presentation of manuscripts.]. . . .

I think this book—the book was written during the winter of 1936–37—of course, with books of this kind, you cannot use lecture notes as anything more than indications of materials to be discussed in the book, for a book cannot be a verbatim transcript of orally given lectures. Any book which would represent a tape-recorded transcript of orally given lectures would be very bad, and vice versa. . . .

Back in Pasadena, I had to think about writing that book, and book writing was assisted by a bad accident. The only two sports I have ever cared for, in fact still care for if I have an opportunity are (a) horseback riding, (b) mountain climbing. So, during winter in Pasadena, I usually went once a week to one of the riding academies where I had a horse and rode him for a couple of hours. That was in the neighborhood of Pasadena. There was a certain trail which led through a gate. This gate led to a highway. I went through this gate many times, but just before that fateful ride, somebody decided to put a rail in the middle of this gate to prevent the access of cars from the highway into the bridle path, a fact which I did not know.

I was galloping, as I did numerous times before. This time, the horse had to pass through half of the gate, rather than the whole gate. So my left knee came against the cement post, and as it proved to be upon X–raying, the patella bone, the knee-cap, was, to quote the surgeon, "broken into four large pieces and some fragments."

Well, needless to say, that had to be followed by an operation. Incidentally, I may say, although the accident itself was not painful, before the operation, bending that broken leg produced one of the most excruciating pains I have ever experienced in my life. So the night before the operation, when I was already in the hospital, my leg, in my sleep—I don't remember whether I had a sedative or not—anyhow, it fell from the bed, and I think the yell which I emitted probably woke up most of the people in the neighboring rooms.

Anyway, the operation was made, and I had to stay in bed for, I believe, something like three weeks, in a cast which enclosed my entire leg. Then for a couple months I had to use a crutch, and several times, at irregular intervals, I had the swelling of the knee. Well, anyway that was the time when I wrote that book.

It was a time when I was tied to a bed, bedridden, unable to walk very far, but of course I was perfectly healthy in other ways. So a hospital bed was rented, where I could be cranked up, and a little wooden table was either bought or made, I forget. Anyhow, I had a very productive time, working most of the day from morning until evening writing. That, of course, would be wrong to make out that the whole book was written in that position. After all, with book writing, for most people, for me at least, what really matters is to write, so to speak, a skeleton, to express, shall we say, an outline of whatever it is that you want to express. Now, the documentation, the references, the details come in relatively easily later. Obviously, you have to go to a library, which I did during the several subsequent months. I don't remember the precise date, but when I sent the manuscript, Columbia Press did a very rapid job on it, so the first edition was in 1937, and in fact early 1937. They did a really very quick job on it. (Dobzhansky 1962–1963, pp. 385, 390, 392–396)

These two oral memoirs are full of information and surmises, some of them contradictory as one might expect from recollections from some thirty years earlier. Reading the oral memoirs again makes me wish that the interviewers (Saul Benison for Dunn and Barbara Land for Dobzhansky) had been much more informed and critical, but we are indeed fortunate to have these memoirs at all.

THE JESUP LECTURES

Many important questions about the new Jesup Lectures arose in this research, but I was unable to answer many of them. Why did the Jesup Lectures and the corresponding Columbia Biological Series disappear for more than thirty-five years, when in fact the lectures and the books published in the series were not only very important but some were still in print? E. B. Wilson's *The Cell* originally appeared in the Biological Series in 1896 and the third edition (1925) was still in print, as was Henry Fairfield Osborn's *From the Greeks to Darwin*, the first book published in the series.

What stimulated Dunn to reestablish the lectures and the Biological Series? I simply do not know the answer to this question. Were H. E. Crampton or J. H. McGregor of the old guard evolutionists in the department nostalgic about the series? Was Franz Schrader, a real force in the department and a very close friend of A. H. Sturtevant, pushing for the new lecture series? Why was the topic "genetics and the origin of species" chosen? (Ernst Mayr suggested this question.) Dunn's view was that the title was chosen at Columbia and Dobzhansky was invited to give the lectures on it, and Dobzhansky's recollection was that he had chosen the topic after being invited to give the Jesup Lectures. Dunn had been a student in the Bussey Institution of Harvard University working primarily with William Castle, and Dunn's expertise was in the physiological genetics of mammals and especially chickens. His background in evolutionary biology was weak, as he admitted in the oral memoir.

Yet Dunn came to the Bussey in Sewall Wright's footsteps and was keenly aware of Wright's seminal work in theoretical population genetics and evolution. Dunn also was important in stimulating and refereeing the debate between Fisher, Wright, and Haldane over the evolution of dominance in the pages of *The American Naturalist* in the late 1920s and early 1930s (for an account and references, see Provine 1986, pp. 243–260). The 1932 International Congress of Genetics in Ithaca, New York, had a popular session featuring Fisher, Haldane, and Wright chaired by Richard Goldschmidt. Dunn, Dobzhansky, and many others were greatly stimulated by that session. As Dunn claims in the oral memoir, he was aware of the work in population genetics but thought it was very theoretical and that laboratory experiments were incapable of discriminating between theoretical hypotheses. Dunn believed that what was needed were ways to relate the debates of theoretical population genetics to the genetics of natural populations.

In his editor's preface to *Genetics and the Origin of Species*, Dunn pointed out that most of Dobzhansky's sources for the book were less than fifteen years old.

> The reasons for this are not far to seek. Variation and heredity had first to be studied for their own sakes and genetics grew up in answer to the interest in these problems and to the need for rigorous methods for testing by experiment all ideas we might hold about them. The requirements of this search drove genetics into the laboratory, along an apparently narrow alley hedged in by culture bottles of Drosophila and other insects, by the breeding cages of captive rodents, and by maize and snapdragons and other plants. Biologists not native to this alley thought sometimes that those who trod along it could not or would not look over the hedge; they admitted that the alley was paved with honest intentions but at its end they thought they could see a red light and a sign "The Gene: Dead End."
>
> That condition, if it ever existed to any marked degree, is again changing, and Professor Dobzhansky's book signalizes very clearly something which can only be called the Back-to-Nature movement. The methods learned in the laboratory are good enough now to be put to the test in the open and applied in that ultimate laboratory of biology, free nature itself. Throughout this book we are reminded that the problems of evolution are given not by academic discussion and speculation, but by the existence of the great variety of living animals and plants. The facts and relationships found in nature have to be examined from many points of view and by the aid of many different methods. Evolution, in the author's words, is a change in the genetic composition of populations, and populations follow laws which may be derived by mathematical reasoning as extrapolations of the known behavior of the fundamental units of reproduction—genes and chromosomes. It is a kind of tour de force that in this book the recent work in this field is fitted into its important place in a way which does not offend the sensibilities of those who are repelled by mathematical formulas. (Dobzhansky 1937, pp. viii–ix)

Dunn of course wrote these words after listening to Dobzhansky's Jesup Lectures and reading his manuscript in its entirety. So one cannot conclude that his views in the preface are those he held at the time he issued the invitation to Dobzhansky to give the lectures, and must be taken as merely suggestive.

Dunn did not know Dobzhansky personally when he asked him to give the lectures. But very plausible (not certain) reasons existed for Dunn to ask him. These relate primarily to the research that Dobzhansky was doing with Sturtevant and the desire of Demerec to attract Dobzhansky to the East Coast.

Demerec and Dobzhansky had known each other for many years. In 1929 Dobzhansky was worried about having a student visa at the same time he was an assistant professor at Caltech, fearing he might be deported back to Russia where he feared severe treatment. Demerec was the person who worked with immigration officials and who persuaded both C. B. Davenport and H. H. Laughlin, the two senior officials at the Station for Experimental Evolution, to write letters supporting Dobzhansky. This episode blew over (the State Department said it was permissible for Dobzhansky to be an assistant professor on a student visa!), but the issue rose anew in 1931. This time Dobzhansky and his wife Natasha left the U.S. for Vancouver in order to return with new professional visas. Once in Vancouver, U.S. immigration officials unexpectedly refused entry to the Dobzhanskys even with new visas. Dobzhansky appealed the refused entry to the Secretary of Labor and wrote a desperate telegram to Demerec informing him of the problem. Demerec immediately sent special-delivery letters to F. D. Ritchey of the Department of Agriculture asking for his intervention with the Secretary of Labor and engineered a night-letter (next-day delivery) writing campaign for renowned geneticists to support the Dobzhanskys. The campaign was successful, and Dobzhansky wrote to Demerec: "I want to thank you for all you did for us. There is no use to say how much we feel obliged to you, because nobody can fail to realize from what a hole you extracted us" (Dobzhansky to Demerec, 13 August 1931). After this, the personal bond plus the great shared interest in *Drosophila* genetics resulted in a voluminous correspondence between Dobzhansky and Demerec.

Dobzhansky was a big supporter of Demerec's proposed Drosophila Information Service (Dobzhansky to Sturtevant, 30 September 1933), and for his part, Demerec offered Dobzhansky several months of research at Cold Spring Harbor with paid travel expenses (Demerec to Dobzhansky, 2 January 1934). Dobzhansky chose instead to gather *Drosophila* in Alaska but expressed a strong interest in coming another year.

In early 1936, when Dobzhansky and Sturtevant were excitedly beginning to sketch out their ambitious plans for research on natural populations of *Drosophila pseudoobscura*, Dobzhansky and Demerec were in constant communication by letter, with Dobzhansky giving Demerec frequent updates. Thus, in late January Dobzhansky wrote: "Sturtevant and myself are gone crazy with the geography of inversions in pseudoobscura, and working on this whole days—he with crosses and myself with the microscope. No short inversions are found. . . . As to our inversions, Mexico seems to be an inexhaustible source of them, and I am beginning to regret that last year only relatively few Mexican strains were collected" (Dobzhansky to Demerec, 26 January 1936).

And three weeks later he wrote: "Sturtevant and myself are spending the whole time studying the inversions in the third chromosome in geographical strains of pseudoobscura. We are constructing phylogenies of these strains, believe it or not. This is the first time in my life that I believe in constructing phylogenies, and I have to eat some of my previous words in this connection. But the thing is so interesting that both Sturtevant and myself are in a state of continuous excitement equal to which we did not experience for a very long time" (Dobzhansky to Demerec, 17 February 1936; for a fuller account of this collaboration, see Provine 1981 and Kohler, this volume).

Demerec at just this time was very keen to develop the genetics department at Cold Spring Harbor in the direction of evolutionary genetics. To this end he had made an offer to Nikolai Timoféeff-Ressovsky, Russian educated but at that time at the Kaiser Wilhelm Institut in Berlin-Buch. He was widely recognized as one of the top evolutionary geneticists in the world at that time; indeed, I have argued that the ambitious plan for the study of the genetics of natural populations hatched by Timoféeff-Ressovsky, Jucci, and Buzzati-Traverso in 1936 was a more robust research plan than that dreamed up by Dobzhansky and Sturtevant at about the same time (Provine 1985). It would have been a great coup to get Timoféeff-Ressovsky, but he did not seem very interested. Just as the hope of attracting Timoféeff-Ressovsky was fading during the spring of 1936, Demerec began to realize that getting Dobzhansky would be even better. The work that Sturtevant and Dobzhansky were doing together was just what Demerec wanted to have at Cold Spring Harbor, and Dobzhansky would give the Station for Experimental Evolution a great boost in the direction of its very name.

On April 20 the invitation from Dunn arrived, and the same day Dobzhansky wrote to Demerec: "Just got a letter from Dunn containing an invitation to stay at Columbia for a few weeks this fall. . . .

Shall reply to Dunn in a day or two, after talking to Dr. Morgan who is away for a few days. Of course, I want very much to go, and hope to be able to do so."

My surmise, unsupported by direct evidence, is that Dunn and Demerec, who were often in contact, conversed about the possible candidates for the Jesup Lectures. Demerec's prime candidate must have been Dobzhansky, and Demerec was prepared to give Dunn chapter and verse about why Dobzhansky was an excellent choice.

Another reason that Dunn was keen to have Dobzhansky was that Dobzhansky was Russian and familiar with an entire body of literature on the genetics of natural populations not generally available in the West, one of Demerec's major aims in attracting Timoféeff-Ressovsky. Thus, in his preface to *Genetics and the Origin of Species*, Dunn wrote: "The author appears not only as geneticist and as student of natural history, but as one who received his training in both fields in Russia. English-speaking biologists have special cause to be grateful for this last fact, for it has enabled Professor Dobzhansky to make available to us many important contributions from workers in the Soviet Union, where researches in this field have been actively prosecuted" (Dobzhansky 1937, p. ix).

John A. Moore, who was a graduate student at Columbia in 1936, suggested another possible reason for Dunn's choice of Dobzhansky (personal communication). Perhaps Sturtevant suggested Dobzhansky as a candidate to his close friend Franz Schrader. Kenneth Cooper (personal communication) says it was commonly believed at Caltech in the late 1930s that when Sturtevant received an offer from the Zoology Department in the mid-1930s, he was offered also the chance to restart the Jesup Lectures but declined the lectures at the same time he declined the position in favor of remaining at Caltech.

One may wonder why Dunn did not choose Sturtevant instead of Dobzhansky. Perhaps Dunn did indeed ask Sturtevant, who refused. One thing to keep clearly in mind is that in 1936 the Jesup Lectures were neither well known nor well funded. Only a handful of biologists at the time knew the lecture series by name. No lectures had been given in more than twenty-five years. The stipend was meager and the living quarters humble. Dunn and others in the Columbia Zoology Department were out on a limb trying to get the series going again. I suspect they had no idea that the series would have such a roaring start or such an outstanding and influential future. Dobzhansky was not made famous because he was invited to participate in a prestigious and important lecture series; he made the Jesup Lectures into a prestigious series from almost nothing.

GENETICS AND THE ORIGIN OF SPECIES

When Dobzhansky gave the Jesup Lectures, they were by all accounts well received. The lectures precipitated a great deal of discussion and criticism. Dobzhansky had many conversations with biologists from Cold Spring Harbor, Columbia, and the American Museum of Natural History. In response to a draft of my earlier essay on Dobzhansky (Provine 1981), Ernst Mayr wrote a comment pertinent here:

> You almost casually mention that Dobzhansky was five months in the East in 1936, but you pay no attention to the impact that his stay in Cold Spring Harbor and New York had on his thinking. I can report from my own experience not only that he visited me repeatedly at the Museum where I demonstrated to him numerous instances of geographic variation and geographic speciation, but also that we talked about this subject by the hour. Also, after each of the Jesup Lectures there were very active discussions that must have had an impact on Dodik. In Cold Spring Harbor, he must have had a considerable input also, and not only from Demerec, Kaufmann, and McClintock. At Columbia there was Schrader, a famous cytologist, and also some botanists, also Gregory, McGregor, etc. The number and extent of influences must have been very great. I realize you couldn't bring this out in the present account, but it simply was a major impact on *Genetics and the Origin of Species* in the crucial year of 1936. (Mayr to Provine, 16 August 1979)

A contingent from Cold Spring Harbor, including Demerec, came for every lecture.

Immediately following the last lecture, in the first week of November, the Dobzhanskys drove across country (by way of Alabama and Florida to visit cytogeneticists Kaufmann and Metz) and arrived back in Pasadena in time for Dobzhansky to write and thank Demerec on 18 November. For more than a month Dobzhansky did nothing on revising his Jesup Lectures. That was because he had received no assurances that the resulting manuscript would be published. On 21 December he wrote to Dunn about the Russian situation in genetics (a very interesting and insightful letter), adding: "Now, today I have received a letter from the Columbia Press, informing me that the Publication Committee decided to publish my prospective opus provided they approve it in its final form (which, I take it, means if they will not discover there something of the nature of pornography, or of crazy or crank ideas). This means that I must sit down to work, which I was not doing during the last month on account of uncertainty. Now, in this connection I have an annoying request to you. Will you permit me to send the MS, or parts thereof, to you before it goes to them, in order that I may have your suggestions on it—the suggestions you have made while I was in New York were good ones, and some more of them would be of great help to me. I realize that I am asking a lot because it will take some of your time of which you have no excess, but I am taking the chance to make this request since you are the sponsor of this business anyway." He added: "There is nothing particularly new or interesting here."

And Dobzhansky set to work writing the book, not ceasing until the manuscript was sent to Dunn. On 12 January he wrote to Demerec: "At present I am a 'literary worker,' that is my time is taken primarily by writing the book." And on 1 February: "I am still engaged in writing and not much else, and this will continue for a month or two. It begins to be tiring, but having wasted so much time on it already, it were bad not to finish the business somehow. The situation here is in general a sad one, somehow this winter every one seems depressed, perhaps owing to weather which is indeed horrible—frost, rain, and above all the 'grunge'—something you are fortunate not to experience ever."

He was taking a break from writing when he rode his horse through the narrowed gate and smashed his kneecap and leg in late February or early March (I have no precise date). On 26 March he informed Demerec: "This is simply to inform you that I am still alive, and gradually trying to rejoin the society of normal men, although it looks that it will take at least a month before I shall be able to walk without crutches. Yesterday I went to the Lab, but it is very hard. . . . My only work is now writing the book, which I hope to finish in a month."

How many authors can hope to finish a book in a month and then do it? On 11 April Dobzhansky reported to Demerec: "Many thanks for your letter of April 2 and the Florida data. They will be included in my chapter on mutations. Incidentally, the book is almost ready, and in two weeks I hope to have it on its way to the publisher. This will be a great relief, and I shall be able to go back to more productive occupations. A full half of a year has been killed for writing the stuff. It remains to be seen whether the product will prove to be worth the efforts."

Exactly on schedule, Dobzhansky sent the manuscript to Dunn and these words to Demerec: "It just so happened that today I sent off the MS of my book to Dunn. . . . You cannot imagine my relief at having finished the book. It was such a lot of work. Now I can again do something with flies. For the last 2½ months it was out of the question" (Dobzhansky to Demerec, 26 April 1936).

Columbia University Press was quick indeed: it began advertising the book in June, and *Genetics and the Origin of Species* was available in mid-October for use in fall classes. Indeed, by modern standards of digital word processing and electronic publishing with digital texts on sophisticated laser printers, it is almost inconceivable for a university press to commission an important book and expect the completed manuscript four months later; or to receive the unedited manuscript and five months later have the book available on the market.

The earlier Columbia Biological Series was beautifully bound with dark rich buckram, with not only the title and name of the series stamped in gold but also an artistic diagram of a cell dividing (no doubt invented for E. B. Wilson's *The Cell*). This diagram, however, had the chromosomes in the division figure backward. Somehow the error was unnoticed and appeared on at least three of the books in the series before someone (I would like to know who, and how this had escaped Wilson's notice) caught the error and put the stamp safely away in favor of an accurate one that subsequently appeared on books in the series. When time came to choose bindings for Dobzhansky's *Genetics and the Origin of Species*, Columbia Press searched and found the old stamps and, sure enough, used the faulty one on the binding. Ernst Mayr delights in telling the story of how he pointed out this error to Dobzhansky, a case of the bird taxonomist telling the working cytologist that the cell division pictured on the spine of his book is ridiculously in error (personal communication).

The book appeared shortly after the State Department approved the citizenship applications of Theodosius and Natalia Dobzhansky. On 10 October, Dobzhansky wrote: "A few days ago we have finally become U.S. citizens. Hope this will be the end of troubles in this sphere. By the way, if your library has a copy of my book, it has it before the author—I do not know what that means. You should get one of the author's copies ..." (Dobzhansky to Demerec, 10 October 1937).

Concluding Remarks

Dobzhansky often admitted that he was a poor historian and keeper even of his own records such as correspondence. His oral memoir is crammed with demonstrably false or highly questionable assertions (but is of course extremely valuable for the historian). Indeed, my inclination is not to trust in detail anything Dobzhansky says in the memoir without further corroboration. The last thing I expected him to have an accurate view about was the accomplishment represented by *Genetics and the Origin of Species*. At Ernst Mayr's conference on the evolutionary synthesis (Mayr and Provine 1980) in May 1974, Dobzhansky was unquestionably regarded by the assembled architects of the synthesis as the person who really started the synthesis with his enormously influential book. The historical evidence for this view is great. But the influence of the book on evolutionary biology should be distinguished from its originality in the field of evolutionary biology at the time. Dobzhansky carefully made this distinction in the oral memoir:

> This book (Genetics and the Origin of Species, 1937) was very well received, I may say. As I look over it now, I see quite a lot of things which are simply naive, and a lot of other things which are plain wrong. ...
> The reason why that book had whatever success it had was that, strange as it may seem, it was the first general book presenting what is nowadays called—George Simpson's preferred expression is, "the synthetic theory of evolution." I prefer to call it "biological theory of evolution."
> Now, by that I certainly don't mean to make a preposterous claim that I have invented the synthetic or biological theory of evolution. It was, so to speak, in the air. People who have to be credited of having, more than anybody, contributed to it, I believe I said already in our interview before were: Fisher, Sewall Wright, J. B. S. Haldane, and in their turn their predecessor was Chetverikov. What that book of mine, however, did was—well, if you wish, in a sense, popularizing this theory. Wright is very hard to read. He has a lot of abstruse, in fact almost esoteric mathematics. Mathematics, incidentally, of a kind which I certainly do not claim to understand. I am not a mathematician at all. My way of reading Sewall Wright's papers, which I still think is perfectly defensible, is to examine the biological assumptions which the man is

making, and to read the conclusions which he arrives at, and hope to goodness that what comes in between is correct. "Papa knows best" is a reasonable assumption, because if the mathematics were incorrect, some mathematician would have found it out.

Now, in this book, of course, I was also able to utilize the Drosophila pseudoobscura and Drosophila persimilis data to good advantage. . . .

I was also able to present in this book a connected story, a more or less connected story, of the basic assumptions which are underlying this theory of evolution. So in spite of the fact that this book certainly cannot in any sense claim originality of the theory, it was the first book which presented it for general biologists. It had very good reviews. (Dobzhansky 1962–1963, pp. 397–400)

If there exists a more accurate brief evaluation of the influence and substance of *Genetics and the Origin of Species*, I am unaware of it.

Acknowledgments

I am greatly indebted to Ernst Mayr, John Moore, and Kenneth Cooper for informative and helpful conversations about this paper. I also wish that it had been possible to interview Dobzhansky, Dunn, Sturtevant, and Demerec. These interviews would have greatly enriched and complicated the historical account of the topic under discussion here. I still think that we should move much faster interviewing elderly scientists. A tremendous amount of historical understanding dies irretrievably with each of them.

References

All correspondence cited in this paper is from the outstanding collection at the American Philosophical Society in Philadelphia. I strongly recommend Glass 1988 as an initial guide to these holdings in the history of genetics and evolution, but I would add that many collections have been added recently, including those of Sewall Wright, Phillip Sheppard, and Arthur J. Cain.

Dobzhansky, Th. 1937. *Genetics and the Origin of Species*. New York: Columbia University Press.

——. 1962–1963. "Reminiscences of Theodosius Dobzhansky." Typed transcript. 2 parts. Oral History Research Office, Columbia University, New York.

Dunn, L. C. 1961. "The Reminiscences of L. C. Dunn." Typed transcript. Oral History Research Office, Columbia University, New York.

Glass, B. 1988. *A Guide to the Genetics Collections of the American Philosoph-ical Society.* American Philosophical Society Library, Publication Num-ber 13. Philadelphia: American Philosophical Society Library.

Mayr, E, and W. B. Provine, eds. 1980. *The Evolutionary Synthesis: Perspec-tives on the Unification of Biology.* Cambridge: Harvard University Press.

Provine, W. B. 1981. "Origins of the 'Genetics of Natural Populations' Series." In R. C. Lewontin, J. A. Moore, W. B. Provine, and B. Wallace, eds., *Dob-zhansky's Genetics of Natural Populations I–XLIII* (New York: Columbia University Press), pp. 5–83.

————. 1985. "The study of the genetics of natural populations during the evolutionary synthesis of the 1930s and 1940s." In L. Bullini, M. Ferra-guti, F. Mondella, and A. Olivero, eds., *La vita e la sua storia* (Sciencia), pp. 121–128.

————. 1986. *Sewall Wright and Evolutionary Biology.* Chicago: University of Chicago Press.

■

Fly Room West: Dobzhansky,
D. pseudoobscura,
and Scientific Practice

Robert E. Kohler

THIS IS A case study in scientific practice in T. H. Morgan's school of *Drosophila* geneticists at the time that Theodosius Dobzhansky was a member of the "fly group" in 1927–1940.[1] Two aspects of *Drosophila* practice especially interest me. First, the production process: I will treat *Drosophila* as a scientific instrument, as a piece of laboratory technology. *Drosophila* was not so much an object of study as a means for producing genetic knowledge. Standard organisms like *Drosophila* can be seen as systems of production, designed artifacts that have skills and procedures built into them through long use. Such instruments have many potential uses; the question is, for what are they actually used, and how are such choices made?

A second aspect of scientific practice is the "moral economy" of production, that is, the social conventions that control access to means of production and that regulate the distribution of credit for achievements—the intellectual wealth produced by research. The rules of citation and coauthorship are familiar examples, but others, more fundamental, remain largely unexamined. How, for example, is the ownership of problems and methods determined in groups like the fly group, where work is done collectively? How much of a group's common skills and problems are disciples permitted to take with them when they leave for independent careers? Every system of scientific production has a moral economy, and the key question is, what is the relation between production and moral economy? How do group values and conventions of behavior shape production, and

---

[1] For sources and further details, see Robert E. Kohler, "Drosophila and Evolutionary Genetics: The Moral Economy of Scientific Practice," *History of Science* vol. 29 (1991), pp. 335–75; and Robert E. Kohler, *Lords of the Fly: Drosophila Genetics and the Experimental Life* (Chicago: University of Chicago Press, forthcoming).

how do changes in production upset the human and moral relations among researchers?

Most histories of the Morgan fly group focus on its first decade or so (1910–1920s), but the late years are to me even more interesting. In the Caltech years (1928–1940s) the group was more complex socially; several distinct generations of workers collaborated and competed. The original program of mapping genes in *D. melanogaster* had been routinized (Morgan thought so, anyway), and the group was striving to identify the big, productive fields of the future. The group also had to deal with growing external competition from other groups of Drosophilists, especially H. J. Muller's and John Patterson's at the University of Texas and Milislav Demerec's at the Carnegie Laboratory at Cold Spring Harbor.

Let us begin with the dramatis personae. The first-generation Drosophilists, Alfred Sturtevant and Calvin Bridges, had quite different roles in the group. Bridges was keeper of the *Drosophila* stocks, the central nerve center of the *Drosophila* exchange network. He was devoted to systematic mapping. Sturtevant was the keeper of *Drosophila* lore, with his vast collection of reprints and comprehensive knowledge of the canon of *Drosophila* genetics. These resources were reflected in their roles within the group. Bridges was a general troubleshooter and technical resource. Sturtevant unofficially took charge of assimilating graduate students and visitors into the fly group, providing them with research projects that meshed their personal skills with the group's current research program.

The principle second-generation Drosophilists were Jack Schultz, George W. Beadle, and Theodosius Dobzhansky. They had quite different interests: Schultz, the physiology of gene expression; Beadle, developmental genetics; Dobzhansky, evolutionary genetics. They were less tied to old lines of work than the first generation and were more open to new systems of production. The younger members of the group, especially Schultz and Dobzhansky, were big talkers and free-wheeling intellectuals. The fly group was the center of communication of the *Drosophila* community, and shop talk was incessant and intense. It is no accident that the two major trends of the 1930s, developmental genetics and population genetics, both took shape at Caltech.

The fly group had a distinctive work-style. It was organized for production. The old fly lab at Columbia had a common work space: no doors, no private offices, and though the group's new laboratory

TABLE 1
Publications and Coauthorship, 1927–40

	Sturtevant	Bridges	Dobzhansky	Schultz	Other	Percent
Sturtevant	31	0	8	1	6	48
Bridges	0	36	2	1	12	41
Dobzhansky	8	2	68	5	10	36
Schultz	1	1	5	18	3	56

at Caltech had separate rooms, the collective work style continued as before. Work was done communally, with constant cross talk. The laboratory was to outsiders a picture of confusion; but to its inhabitants, it was a place where work could be done. Production had the highest value, and human relationships were designed to that end. It was often impossible to say with whom new ideas originated, and for that reason credit was given to those who did the work. Students were regarded as members of the group, not the personal property of individual professors. There were few technicians, and everybody from Morgan on down did their own work at the bench.

Collaborations were the key to practice. During the time that Dobzhansky was in the group, roughly half of all publications of the group were coauthored: from thirty-six percent for Dobzhansky to fifty-six percent for Jack Schultz (Table 1). These were not casual or fictive collaborations but formative personal interactions. Schultz was instrumental in Beadle's invention of developmental genetics. It was Schultz and Dobzhansky's collaboration on sex-determining genes in the early 1930s that led Dobzhansky to study wild populations of *D. pseudoobscura*. Similarly, Sturtevant and Dobzhansky's collaboration on the phylogeny of local races of *pseudoobscura* led directly to Dobzhansky's invention of population genetics. Cooperation was not just policy, and it was never preached. Rather it was implicit in the production practices of the fly group, ensured by Morgan's unique authority as founder of the field. (As Morgan's former students, even Sturtevant and Bridges were reflexively deferential to "the Boss.")

This remarkable collective style of production was not without its costs, however. One such cost was a restriction of the personal ambition of the core group. The implicit moral economy discouraged big individual projects except as part of the group program. (Sturtevant began a big project on comparative genetics in the early 1920s but gave it up. Virtuoso puzzle-solving proved more in tune with the group's productive and moral economy.) Another cost was the culture of scarcity that Morgan cultivated. His notorious stinginess with resources, though usually seen as a personal quirk, was more likely meant to keep everyone in the group on a more or less equal level. In a culture of chronic scarcity it was more difficult for personal fiefdoms and cliques to form. Loyalty to the group overrode desire for independent careers, especially among first-generation Drosophilists. Sturtevant, had he wanted to, could have been the boss of any genetics department in the world, yet he stayed with Morgan. He benefited from being at the center of the *Drosophila* world, of course, but he paid a high price in the lack of administrative power and of opportunities to develop his own independent research program.

Dobzhansky became aware of this other side of the moral economy of the fly group soon after his arrival at Columbia in 1927. There was constant and bitter complaining about Morgan, who was said to have taken all the credit for achievements of the whole group. Sturtevant took Dobzhansky aside and unloaded his grievances about limited resources and restricted career. (He liked to quote a colleague's quip that he, Sturtevant, was Morgan's greatest discovery.) Dobzhansky was terribly shocked at this distinctly unheroic behavior of people whom he had idolized from afar as scientific heroes. But such behavior was not in fact human weakness. Conflict was the price the group paid for its highly productive collective work style and the self-denial it entailed. Grumbling was a way of coping with these structural conflicts without tearing the productive group apart. The alternatives, after all, were to challenge Morgan's leadership or to leave.

Dobzhansky was, of course, completely unaware of this deeper meaning of the group's collective behavior. Both as a foreign visitor and as a second-generation Drosophilist he was not subject to the same constraints, not at first anyway. His personal ambition was less constrained by loyalty to the group and to older traditions of group research. Sturtevant, as was the custom, took charge of assimilating Dobzhansky into the group's practice. He set him a problem growing out of his own research, which, had it worked out, would have neatly enabled Dobzhansky to apply his great skill in dissection and morphology to a central aspect of the group's program. The project did not work, however, and Dobzhansky's full initiation into the group came about more as a result of his own independent initiative in learning Muller's new technique of X-ray mutation and applying it with stunning success to the fashionable problem of translocations.

By 1933 Dobzhansky was a fully assimilated member of the fly group: an assistant professor, close collaborator with Sturtevant and Schultz. As a talker and intellectual gadfly he learned to hold his own even with Schultz. He became part of the moral economy of the fly group, complaining as loudly as anyone about Morgan's stinginess and about his own need for more resources for his expanding research. A fast worker and compulsive publisher, Dobzhansky produced papers twice as fast as Sturtevant and four times as fast as Schultz (see Table 1). He also, however, was taught the difference between the moral rights of first- and second-generation Drosophilists to complain, when one day Sturtevant icily ordered him never again to criticize Morgan in his presence. Dobzhansky was dumbfounded. How could his best friend blame him for doing what he himself did all the time? The meaning of this episode, which Dobzhansky never grasped, was that Sturtevant had earned the right to complain through self-denial, Dobzhansky had not. Second-generation Drosophilists could give their ambitions freer rein and were less bound to defer to the moral economy of self-restraint.

Such conflicts over the allocation of resources and authority worsened when Dobzhansky's remarkable research on *pseudoobscura* became a bigger and more important part of the collective practices of the fly group.

How Dobzhansky came to form his "lifelong friendship" with *D. pseudoobscura* is a long and complex story, which has been well told by William Provine. Suffice to say here that it evolved out of research with Schultz in 1930–1932 on hidden "sterility" genes that determined hybrid sterility (an old interest of Dobzhansky's because of its possible role in the process of speciation). When hybrids of *D. melanogaster* and *D. simulans* proved inconvenient for large-scale experimentation, Dobzhansky and Schultz turned to hybrids of the more closely related A and B races of *D. pseudoobscura*. Apparently they were planning a big project on hybrid sterility when the unexpected genetic variability of *pseudoobscura* diverted Dobzhansky from his work with Schultz to the genetics of wild populations. Accustomed to working with *melanogaster*, a thoroughly domesticated and "standardized" laboratory creature, Dobzhansky and Schultz had not expected that stocks of *D. pseudoobscura* collected from the wild would have such marked genetic differences. But *D. pseudoobscura* turned out to be a wild creature, quite unlike its domesticated cousin, and its introduction into the laboratory transformed the practices of the fly group.

But why did Dobzhansky go to the wild for *pseudoobscura* and thus unwittingly inject an abundant natural variability into the domestic laboratory scene? The answer is simple: *pseudoobscura* was not available on the *Drosophila* exchange network. (A few stocks were available from Donald Lancefield's earlier work, but not enough.) Unlike *melanogaster, pseudoobscura* was not a scavenger and camp follower of humankind, preferring wild places. So Dobzhansky was forced to go afield to collect wild stocks. The point is that the *Drosophila* exchange network traded only in "standard" domesticated strains. It had been designed to enable researchers to work with the same standard organisms and thus to do comparable experiments without having to worry about genetic variability. For most genetic work the absence of variability had become a necessity. Thus, when Dobzhansky brought back wild *pseudoobscura* from the field he introduced a wild and potentially transforming element into standard laboratory practice. *Pseudoobscura* became a new instrument, and a new mode of production developed around it. This new system of production, which combined elements of field and laboratory, altered the productive and moral economy of the fly group, elevating Dobzhansky to senior status and transforming his relations with Schultz and Sturtevant.

At first (1932–1935) Dobzhansky worked mainly with Schultz on *pseudoobscura*, continuing their earlier work on hybrid sterility. As Dobzhansky's interest in geographical variability grew, however, he ended his collaboration with Schultz and revived his earlier collaboration with Sturtevant. Why did Schultz drop out? Probably he did not like the new wild organism and its wild variability. Also, unlike Dobzhansky and Sturtevant, Schultz had little interest in problems of evolution (his interests ran more to physiology and biochemistry). He preferred the familiar, constructed "standard" system of *D. melanogaster*. Dobzhansky later recalled how skeptical his colleagues were at first about *pseudoobscura*. Why work with a new organism that had no genetic markers, no neat trick chromosomes, no network of exchange, no fund of craft knowledge on which to draw? Why reinvent the wheel with a new organism? Thus, Schultz stuck to the problem of sex determination in *melanogaster* while Dobzhansky, with Sturtevant, moved on to fabricate *pseudoobscura* into an instrument of evolutionary genetics. By 1934 Dobzhansky had 300 stocks of *pseudoobscura* in cultivation and had abandoned *melanogaster* for good.

Dobzhansky and Sturtevant's 1934–1936 collaboration was extraordinarily intense and stimulating to both men. At first Sturtevant's vision of evolutionary genetics predominated. Sturtevant had been interested in phylogenetic relations of *Drosophila* since the early 1920s, and the numerous overlapping genetic inversions of geographical races of *pseudoobscura* made it possible for the first time to infer historical relationships. Although Dobzhansky had never been much interested in phylogeny, the new method was just too sensational not to follow Sturtevant's lead. Their intense and emotional partnership was extraordinarily productive, but it also contained the seeds of conflict. Both Dobzhansky and Sturtevant were planning grand long-term projects and saw *pseudoobscura* as the vehicle of a greater or revived career. But they envisioned evolutionary genetics in different ways, and by 1937 differences muted by the thrill of discovery were coming more into the open.

Dobzhansky's program of research with *pseudooscura* was initially very broad, encompassing genetics, cytology, embryology, physiology, and sex determination (even some biochemistry and ecology, which Dobzhansky never fancied). His plan, clearly, was to fashion a wild organism into a laboratory instrument. Soon, however, Dobzhansky became completely absorbed in analyzing the variability of regional populations. This was a novel mode of practice. Field and laboratory practices were united, and natural variability had been transformed from a technical difficulty into a most valuable resource. This new mode of practice had an immense appeal to Dobzhansky precisely because it did bring field and lab together. Dobzhansky had always loved camping and traveling to wild places. *Pseudoobscura* gave him the opportunity to bring together these two sides of his life, work and leisure. Working in the laboratory with wild material also enabled him to work in the way he liked best. Dobzhansky was always impatient to produce lots of data fast, and screening wild stocks of *pseudoobscura* was not as slow and exacting as classical genetic and cytological work. Genetic analysis of natural populations did involve much routine screening, of course, but the cytology was quick and dirty, quite different from the meticulous and tedious cytology of traditional laboratory practice. Analysis of wild stocks was fast paced. It was intense and exciting. It produced vast amounts of new and interesting information in very short periods of time, unlike the more exacting work with standard organisms. This novel mode of production, uniting field and laboratory, was Dobzhansky's most important creation, and it was the foundation of his emerging vision of an evolutionary genetics of natural populations.

Sturtevant's program for *pseudoobscura* was equally large and ambitious but quite different. Sturtevant envisioned a comparative genetics of all the *Drosophila* species of the world: a systematic multi-species comparison of homologous mutations. In 1937 he was planning to seek a grant of $10,000 a year for a big long-term project. By 1939 he had forty-two species under cultivation and was at work comparing and tabulating twenty-seven standard characters, searching for clusters and clues to phylogenetic relationships. All this was highly uncharacteristic behavior for Sturtevant, who was known for his dislike of big projects. Dobzhansky was surprised and skeptical about Sturtevant's traditional approach to evolution. As a latecomer to the fly group, Dobzhansky could not have understood just how important *pseudoobscura* was to Sturtevant's career. The project was a revival of his abandoned project in the early 1920s on comparative genetics. It marked the end of a period in which Sturtevant had seemed to lose his way, doing many striking little things but no big thing. For Sturtevant, *pseudoobscura* was the biggest thing since Morgan's original program of gene mapping: it was to be the rejuvenation of the fly group and of his own career.

Dobzhansky's and Sturtevant's designs for *pseudoobscura* were rooted in different kinds of practice. Sturtevant was interested in phylogenetic relationships and used methods of classical laboratory genetics. Dobzhansky was interested in the dynamics of change in local populations and combined laboratory and field practices. The fundamental difference between the two men lay in their different experimental work styles—that is, in the way they used *Drosophila* to produce knowledge. Sturtevant used *pseudoobscura* in the same way he did the domesticated *melanogaster*. Standard species types were essential for comparative genetics. He constructed homozygous stocks, then introduced sections of chromosomes. It was classical genetic practice, highly precise but so laborious that relatively few experiments could be done. Dobzhansky, in contrast, used *pseudoobscura* in a new way, in which the natural variability of wild strains became the key resource. Instead of precise genetic analysis, he used a quicker, biometric analysis of the morphology of testes. For cytological assay he made quick, disposable smears, avoiding the more laborious permanent preparations of classical practice. Dobzhansky's new mode of practice seemed sloppy in the eyes of classical geneticists. It produced quickly the vast amounts of data required for populational analysis, but was it reliable, Dobzhansky's colleagues wondered. Uncertainty over the criteria of reliable practice was symptomatic of a fundamental change in the way *Drosophila* was used.

It was a conflict not of good and bad practice but of different practices, new and old. Sturtevant tried to domesticate *pseudoobscura*, extending traditional laboratory practices and standards to a wild creature, controlling and confining its variability. Dobzhansky brought a wild creature into the laboratory, multiplying and exploiting its variability and transforming traditional laboratory practice. Dobzhansky and Sturtevant made *pseudoobscura* into two different kinds of scientific instruments, two different modes of scientific practice, embodying different conceptions of the relation between laboratory and field.

Dobzhansky's novel work style, with its unexpected and quite astonishing productivity, resulted in a serious rift with Sturtevant. This rift was revealed to Dobzhansky in a dramatic and unsettling way in May 1936. Briefly, the story is this. Dobzhansky had received a very flattering offer from the University of Texas. He talked it over with Sturtevant, his closest friend, who refused to tell him what to do but made it clear it would be in his interest to accept. Milislav Demerec advised him more bluntly that Caltech could never give his work on population genetics the support it deserved and that for the sake of his career it was time to get out from under Morgan's protective wing. Demerec also, however, held out the prospect of an offer from Cold Spring Harbor, where Dobzhansky really wanted to be. So he accepted Texas, anticipating a subsequent move to Cold Spring Harbor. But did it really make sense to move twice? Dobzhansky changed his mind and wrote Patterson declining the Texas offer. Morgan was delighted, but not Sturtevant. He said nothing when Dobzhansky told him, but the dismayed expression on his face revealed all. Sturtevant did not want him to stay—Sturtevant, his closest friend and collaborator. Dobzhansky was stunned. What was going on? This revelation, occurring at the height of their collaboration, marked the beginning of a personal rift between the two men, which grew steadily wider and more bitter.

What *was* going on? Dobzhansky never really understood, and Sturtevant's views died with him, as he refused ever to discuss the matter. Their falling out has been a taboo subject among older geneticists, who saw it as an unseemly lapse by heroes of their discipline. In fact, the rift between Dobzhansky and Sturtevant offers a fascinating glimpse of the productive and moral economy of the fly group in its later years.

What happened, I think, was this. The astonishing productivity of *pseudoobscura*, and the unexpected divergence in practice that it entailed, upset the moral economy of the group. *Pseudoobscura* became too big for the group. Bridges, when he died in 1938, was the last member of the group working with *melanogaster*. Dobzhansky became too big for the group, as the growing fame of his work on natural populations put him, a junior member, into direct competition with Sturtevant for the leadership of the fly group.

To understand the intensity of this rift we need more context. The crucial element from 1936 to 1940 was Morgan's impending retirement. In September 1936 Morgan turned seventy, the customary retirement age, but he could not bring himself to retire and did not do so officially until 1940. These four years were marked by increasing uncertainty and personal tensions within the fly group. Everyone was on edge. Who was going to succeed Morgan as chief? Sturtevant was the obvious person, but why was Morgan hesitating? Would Morgan's retirement mean the termination of the Carnegie Institution's annual grant? (The fear was that the Institution would shift its funds to its own genetics laboratory at Cold Spring Harbor.) No doubt "the Boss" enjoyed exercising authority and was reluctant to give it up. It is also clear, however, that Morgan had good reasons to delay. The new head of the Carnegie Institution, Vannevar Bush, was in fact eager to terminate the group's grant. Also, Morgan had reason to doubt that Sturtevant would be an effective head of Caltech's Biology Division. Sturtevant, like Bridges, had been a long-term research associate paid by the Carnegie grant and had no academic appointment until 1928. He had had no administrative experience, and Morgan's habit of command, making every financial or political decision himself, had given Sturtevant no opportunity to learn. Sturtevant had become unable to make practical decisions, as Morgan pointedly observed to Caltech president Robert Millikan when the question of the succession came up.

Sturtevant was in a most uncomfortable position from 1936 to 1940. He expected to be offered Morgan's position, but he was also aware that it was the last chance he would have for a new and independent career somewhere else. He was torn between loyalty to the fly group and the desire for the independence he had never permitted himself to grasp. Thus the inherent moral conflicts of group practice surfaced as Morgan's impending retirement relaxed restraints on individual ambitions. It had always been Morgan's authority, and the deference he uniquely could command, that held the moral and productive economy of the fly group together. Without Morgan, how would authority and resources be allocated among the members of the fly group? Would the group have a central program? How would ambitious and expansive individual projects like Dobzhansky's be contained?

The rivalry between Sturtevant and Dobzhansky was not personal but structural, and it was experienced in different ways by the two friends. Sturtevant, though he was far too much the gentleman ever to talk about it, was probably aware of the potential for rivalry. Sturtevant's uncharacteristic plunge into a big and expensive research project was, I believe, a self-conscious effort to establish his vision of evolutionary genetics as the new research program of the fly group, just as Morgan had with gene mapping twenty-five years earlier. Sturtevant seemed to be trying, late in the game, to be acting like the head of a premier program of experimental biology. Dobzhansky, in contrast, seems to have been blissfully unaware of the potential competition. His program in the population genetics of *pseudoobscura* was driven by a desire to advance his career, of course, but his eye was not on Caltech. Sturtevant, with his greater experience and sensitivity to the symbolic meaning of behavior, was sensitized to the issue of competition well before Dobzhansky was—hence Sturtevant's dismay when Dobzhansky decided not to go to Texas, and Dobzhansky's horrified surprise at his friend's involuntary revelation of his true feelings.

Dobzhansky and Sturtevant's remarkable work on natural populations of *pseudoobscura* had in fact made Dobzhansky Sturtevant's equal in the world of Drosophilists. It was a meteoric ascent. Morgan had promoted Dobzhansky to full professor as an inducement to refuse the Texas offer. The Texas position, and the position at Cold

Spring Harbor that Demerec was trying to arrange for Dobzhansky, were ones that Sturtevant would have been proud to have been offered. Most important, perhaps, was L. C. Dunn's invitation to Dobzhansky to revive the prestigious Jesup Lectures on evolution and genetics at Columbia in 1937. Imagine what Sturtevant must have felt when not he but his junior colleague was invited to present their joint work to the world and to publish it in book form under his name alone. What a blow to a man who hoped to put his own personal stamp on the emerging field of evolutionary genetics. Sturtevant had every reason to be jealous, and given his gentlemanly ideals and belief in reason, experiencing such feelings must have been an unpleasant experience. No wonder he never talked about it.

The point is this: the remarkable productivity and novelty of Dobzhansky's new style of work with *pseudoobscura*, in the context of Morgan's constantly delayed retirement, disrupted the moral economy of the fly group. There was no one clearly in charge, no clearly defined group agenda. Dobzhansky's work on natural populations of *pseudoobscura* was outgrowing customary limits on individual projects. Dobzhansky was outgrowing his role as a second-generation Drosophilist. Habits of self-restraint, which had previously been respected out of deference to Morgan, were breaking down.

In the late 1930s, without quite realizing it, Dobzhansky broke most of the rules of the group's moral economy. He became more aggressive in recruiting graduate students to work on his project, for example. He had always wanted more students, and the increasing need for hands to process the deluge of new material from the field made the need more urgent. Returning from a two-year leave in 1939, Jack Schultz remarked upon the striking change in the way graduate students were treated in the group. Before, they were independent apprentice members of the group as a whole. Now they seemed to be regarded as hands for individual professors, especially by Dobzhansky. Schultz did not like this new competitive spirit, and most graduate students did not either, preferring to work in the old way with Sturtevant.

So, too, with the distribution of research funds. Dobzhansky's grumbling about Morgan's stinginess with research money grew more strident, and with some reason: field work was more expensive than laboratory work. It cost money to travel, even when Dobzhansky and his fellow pseudoobscurists traveled on the cheap, camping out and renting cabins once a week to bathe. As Dobzhansky's collecting trips extended beyond California to Alaska and Central America, his needs far outgrew the group's internal resources and he was obliged to learn the delicate, and to him quite mysterious, art of grantsmanship. Accustomed to a European system that operated by personal networks, Dobzhansky found it difficult (despite constant coaching by Demerec) to understand the more meritocratic and entrepreneurial American system of foundation grants. Nor could he fathom why Morgan was putting so much money into biochemistry and plant physiology instead of genetics. As head of the Biology Division, Morgan had many interests to satisfy, but Dobzhansky did not understand the politics of American university departments. He could only assume that Morgan personally thought the biochemists were more worthy of support than he was—hence his increasingly bitter complaints about Morgan and his increasingly aggressive claims for a larger share of the group's resources.

The late 1930s were not a happy time for the fly group and especially not for Dobzhansky. His collaboration with Sturtevant was over by 1937, to the lasting regret of both. They remained friends (though now in separate rooms), but warmth and trust gave way to a cool and wary relationship. Torn by conflicting emotions, Sturtevant was often withdrawn and irritable. Scientific arguments took on an unpleasant personal edge. Dobzhansky became a detached and disheartened observer of the Caltech group in disarray. He was increasingly eager to leave and did not hesitate to accept the offer from Columbia in 1940. The only things he would miss, he wrote Demerec, were the California mountains and deserts which he loved so much and which had become so integral a part of his new mode of scientific practice.

The breakup of the productive and moral economies of the fly group was not personal but structural and generational. Dobzhansky was not alone in his aberration. Schultz, too, was caught up in a bitter dispute with Morgan and Sturtevant after Bridge's death left Schultz saddled with the responsibilities of routine stock-keeping. A similar conflict was being played out at just that time in Caltech's Chemistry Division, where Linus Pauling was challenging the leadership of his aging mentor, Arthur A. Noyes, by aggressively pressing for more institutional resources for chemistry. It was the same pattern: a hard-driving and productive second-generation star who, lacking the experience of the founding generation and their habits of self-denial, perceived the moral economy of group practice in a more individualistic way.

What, in conclusion, does Dobzhansky's experience in the fly group reveal about the collective scientific practices? It highlights, for one thing, the preeminence of production in science. Dobzhansky's career as a Drosophilist, from his early work on pleiotropic genes through hybrid sterility to natural populations, is best understood as a succession of different ways of using *Drosophila* as an instrument to produce new knowledge—successive modes of production. The way in which Dobzhansky and Sturtevant fashioned *pseudoobscura* into different kinds of instruments reveals an unexpectedly complex dynamic between field and laboratory practices. Follow evolving instruments, and they will reveal fundamental historical continuities and transitions.

Dobzhansky's experiences in the fly group also reveal the importance of the human relationships or moral economy of group practice. Production and moral economy are the two sides of group research; human relationships shape the choice of instruments and problems, and vice versa. Sudden changes in productivity or production methods inevitably alter the moral economy of research groups. It is no accident that the rift between Dobzhansky and Sturtevant occurred precisely at the most productive and innovative point in their collaboration. The direction that innovations take is shaped by the nature of the production process. Dobzhansky's leap from hybrid sterility to the genetics of natural populations depended on his collaborations with Schultz and Sturtevant—even, perhaps, on his estrangement from Sturtevant. Is it not possible that intragroup conflicts, by isolating Dobzhansky from established laboratory practices, made possible the rapid evolutionary divergence of a new species of scientific practice?

Dobzhansky on Evolutionary Dynamics: Some Questions about His Russian Background

Richard M. Burian

THIS PAPER poses some questions regarding the influence of Dob-zhansky's Russian background on his later views regarding evolutionary dynamics. Since I know Russian materials only at second hand, I cannot seriously answer the questions raised here; my main hope is to provoke a response from those who are familiar with, or who are able to investigate, the Russian roots of Dobzhansky's work. This said, I shall focus mainly on Dobzhansky's response to the views of Iurii Filipchenko,[1] Dobzhansky's close friend and mentor[2] and the person who coined the term *macroevolution* (Philiptschenko 1927).[3]

I suspect that Dobzhansky's arguments for the continuity of evolutionary mechanisms at all levels of the systematic hierarchy were shaped in part by his concern to overcome Filipchenko's objections to the (neo-) Darwinian extrapolation from micro- to macro-evolution. Furthermore, and this cannot have come easily, in taking this position Dobzhansky sided with Chetverikov and the Moscow school

[1] See Krementsov, this volume, for a discussion of another component of the Russian background of great importance in shaping Dobzhansky's views on species and speciation, namely his involvement with the debates on these topics among various Russian entomologists during his formative years. See also Adams (1988, 1990c, 1990f, and in preparation); and Alexandrov, this volume, for further background on Filip-chenko, his views on macroevolution, and his relations with Dobzhansky.

[2] A major stimulus for the meeting leading to the present volume was the examination of the rich correspondence between Dobzhansky and Filipchenko, the bulk of which occupies the period after the former's departure for the U.S. in 1927 until the latter's death of meningitis in 1930. That correspondence, which will be available in the Library of the American Philosophical Society, will be published by Mark Adams in due course. It should go a long way toward clarifying the issues raised here.

[3] See Adams 1988, 1990f, and in preparation, to which I owe the citation; and Alex-androv, this volume. Many of the issues I raise regarding the relations of Dobzhansky and Filipchenko have been raised by Adams's work in its numerous variants.

(and various Russian entomologists) against Filipchenko. Thus, the investigations that I wish to stimulate concern the role that Dobzhansky's Russian background played in the development of his views on evolutionary dynamics during the period from about 1925 to about 1950.

I shall take for granted that Dobzhansky participated in what Gould (1983) has called "the hardening of the evolutionary synthesis." This point is hardly original (see, e.g., Provine 1986, chapters 10–12) and is probably no longer controversial. What is partly new, in English at least, are the suggestions, pursued here, that Dobzhansky's increasing adaptationism in microevolutionary theory needs to be interpreted against the background of earlier disputes in Russia as well as the more familiar ones in the United States and England, and that it marks a major departure from the Russian biological traditions within which Dobzhansky was raised.

Those traditions, although partly Darwinian, are well represented by Leo Berg, Filipchenko, various orthogenetic theories, and above all, the belief that evolution at or above the species level cannot be fully accounted for by intraspecific competition and its consequences.[4] Particularly relevant to the issues at hand are Dobzhansky's increasing conviction, from about 1940 on, that *intraspecific* differences among organisms and populations, although perhaps brought forth by historical accident or genetic drift, are typically maintained by selection and by *intraspecific* competition, and his ever-stronger insistence that no macroevolutionary mechanisms other than the ones described in microevolutionary theory are required (or play an important role) in nature. Now if selection maintains the differences among organisms that cause populations to differentiate, and the differences among populations are converted to differences between species by a continuation of the very same selective processes, distinctions between micro- and macroevolution become distinctions only of scale; no distinct macroevolutionary processes need be invoked. This extrapolationist view, parallel to Darwin's, is characteristic of the so-called synthetic theory of evolution, which Dobzhansky helped create. I take it to run against the grain of much of the Russian evolutionary biology with which he was familiar, including evolutionary theory as it was taught and practiced

[4] For relevant background regarding the anti-Malthusian tendencies of Russian Darwinism and anti-Darwinism, see Todes 1989.

in St. Petersburg (Petrograd, Leningrad). It is Dobzhansky's growing confidence in this selectionist and extrapolationist position during the 1940s and 1950s at which my questions are directed.

STAGE SETTING

What do I assume about Dobzhansky's Russian background? I assume that, unlike many of the American geneticists who were his contemporaries, his primary training was not as a breeder or as an experimentalist but as a natural historian, entomologist, and field biologist (Gould 1982). I assume that he acquired his deep interest in fundamental questions of evolution from his earliest days as a biologist and that he was not taught, as were his American colleagues, that evolutionary questions are hopelessly speculative. In any case, he clearly would not have been satisfied to work in genetics or in any other discipline unless he could bring his work to bear on evolutionary problems—including Darwin's fundamental problem concerning the origin of species and the question, much debated in the Russian and the American settings, of the causes of variation among organisms. From this perspective, perhaps by 1927 and surely by the time he had been in Morgan's laboratory for a few years, Dobzhansky had an advantage that few American geneticists of his generation enjoyed: he was in a sound position to bring both field and laboratory studies of genetic variation in natural populations to bear on evolutionary issues.

I also assume that accounts of Russian Darwinism such as those offered by Adams (1968, 1979, 1980, 1988, and 1990a–e), Dobzhansky (1980), Todes (1989), and others are reliable. Thus I take it that the inheritance of acquired characters remained an open question and that a majority of important Russian evolutionists from Beketov and Korzhinskii to Filipchenko resisted Darwin's reliance on the Malthusian mechanism of intraspecific competition. Many of those from whom Dobzhansky first learned evolutionary biology denied the importance of intraspecific competition as a major cause of evolutionary change. They did not deny the importance of competition altogether, but they conceived of it mainly as *inter*-specific competition or as "competition with the environment."

For present purposes, the views of Filipchenko and Serebrovskii appear particularly interesting, though my knowledge of their work rests mainly on secondary literature (e.g., Adams 1990c and 1990e;

Gaissinovitch 1980). To put it briefly, Filipchenko held a form of autogenesis from before his close association with Dobzhansky (which began in 1924) until his death in 1930, denying that within-group variation is comparable with, or able to explain, the differences among higher taxa. Thus he distinguished sharply, as Goldschmidt would later, between micro- and macroevolution,[5] denying that selection acting on within-group variation could explain macroevolution. Genetics could solve the problems of micro- but not macroevolution. Serebrovskii, who had worked in Chetverikov's group in the Kol'tsov Institute, held, in contrast, that variation within populations, though perhaps partly random in character, provided the basis for a full theory of evolution. "All evolution is in essence . . . the evolution of the gene fund" (Serebrovskii 1928, quoted in Adams 1990e).

It is worth noting that Dobzhansky's early allegiance was with Filipchenko, not with Serebrovskii.[6] Yet it was Serebrovskii who originally coined the term "gene fund" (*genofond*, Serebrovskii 1926) and Serebrovskii's use of that term that appears to have led Dobzhansky, by a long and ironic pathway, to coin the English term "gene pool" in 1950 (Dobzhansky 1950; Adams 1979). I turn to an exploration of some aspects of that path.

FILIPCHENKO'S *VARIABILITAT UND VARIATION*

I suspect, but cannot document, that the evolutionary issues that most concerned Dobzhansky when he left Russia in 1927 were (1) the causes of variation within populations and (2) the relation between micro- and macroevolution. At the very least, he was deeply familiar with them; they were central to Russian evolutionary biology. They can be united by a bridging question which played an important role in Dobzhansky's early work (bibliography in Lewontin et al. 1981): are the causes of variation *within* populations the same as the causes of variation *between* populations, between species, and between higher taxa? Already in Dobzhansky's early field studies, as in parallel work by Chetverikov and his school, material was gathered to address this question.

---

[5] E.g., Goldschmidt 1940. As Adams points out, Goldschmidt's use of the terms *micro-* and *macroevolution* derives from Dobzhansky, who got them from Filipchenko. See Note 3.

[6] Additionally, there was the influence of Berg's *Nomogenesis*. See Dobzhansky 1980, p. 233

My suspicion is heightened by a reading of one of the few works of Filipchenko accessible to me, his *Variabilität und Variation* (Philiptschenko 1927). In it, Filipchenko drew several fundamental distinctions. Three are crucial for my argument, for they help explain the structure of Dobzhansky's 1937 book. The distinctions are (1) variation as a condition [*Variabilität*] versus variation as a process [*Variation*]; (2) individual variation versus group variation; and (3) statics versus dynamics. For Filipchenko, these distinctions may be applied only at or below the level of Linnean species, i.e., to individuals, pure lines (biotypes), Jordanons (races or subspecies), or species. "In this regard we go no further . . . , that is [not] beyond the species, for we hold that one must understand by variability [*Veränderlichkeit*] the unlikeness or diversity of individuals or groups of individuals within the boundaries of a species, so that species boundaries are at the same time the natural boundaries of variation" (1927, p. 15).[7]

These distinctions divide the study of variation in relation to evolution as follows. Statics studies variation in individuals (norms of reaction, responses to environmental variation) and groups (differences among groups in the distributions of characters of all sorts, including their extremes and means) over specified ranges of environments. Since statics concerns variation as a condition, its results do not turn on the heritability of that variation and hence do not reveal its evolutionary significance. Heritability, together with its consequences in relation to selection, population size, etc., is studied in dynamics. Study of the dynamics of variation within species requires entirely different methods than those of statics, including the methods of what we would now label "population genetics." This is the proper domain of the familiar laws of Mendelian-Morganian genetics, of Vavilov's law of homologous series, and of their elaborations for the dynamics of populations.

Near the end of the book, Filipchenko returns to the relation between his account of variation and the problem of macroevolution. He affirms that small "stepwise" mutations can accumulate to transform Jordanons and that the Jordanons that form a Linnean species, geographically separated, can be transformed to yield distinct species. Nonetheless, he goes on to reinforce some arguments against Darwin's principle of divergence of characters, stating his own posi-

---

[7] Free translations from this source by RMB.

tion as follows: "It seems to us highly likely that the origin of the characters [that differentiate the] higher systematic categories [requires] some other factors than does the origin of the lower taxonomic units" (p. 91). Two pages later he speculates, with Boveri, Conklin, and Loeb, and against Morgan, that "generic characters, like those of the first developmental stages of the egg, and also those of the higher systematic categories, are determined, not by the nucleus, but by the cytoplasm" (p. 93). Repeating the distinction between micro- and macroevolution, he insists that "contemporary genetics removes the veil from the evolution of biotypes, Jordanons, and Linneons" but not from the evolution of higher systematic categories (p. 93).

THE "PHYSIOLOGY OF POPULATIONS"

In order to keep the present analysis within bounds, I shall concentrate on certain features of the first and third editions of *Genetics and the Origin of Species*, drawing also on the series of papers on "Genetics of Natural Populations" (GNP). The analysis has two purposes: to show that the structure of the first edition of his magisterial book (Dobzhansky 1937) is helpfully understood as a direct response to Filipchenko and to illustrate some of the ways in which Dobzhansky's vision of the dynamics of evolutionary processes changed as this focus disappeared—as it had by the third edition (Dobzhansky 1951).

To appreciate these points, start with the organization of chapters in the first edition, lost in the third. In the first edition (p. 14), but not the third, Dobzhansky says he will treat evolutionary statics in chapters 2–4 and dynamics in chapters 5–10. Statics examines "whether the differences between forms encountered in nature can be resolved into the elements whose origin is known in experiments" (p. 14). Dynamics, in contrast, deals with "the evolutionary process in the strict sense" (p. 14)—the "physiology of populations" and the fixation of differences among populations.

This structure connects intimately to the position Dobzhansky takes on the extrapolation of microevolutionary processes to handle macroevolutionary problems. Near the beginning of the first edition (1937), he laments our inability to understand "the mechanisms of macroevolutionary changes, which require time on a geological scale," adding, notoriously, that "we are compelled at the present level of knowledge reluctantly to put a sign of equality between the

mechanisms of macro- and microevolution" (1937, p. 12). This reluc-
tance, I believe, is not feigned or coy; among other things, Dobzhansky took Berg's, Filipchenko's, and others' reservations about
Darwinian extrapolation seriously. Thus, although the equality of
mechanisms is maintained consistently throughout the first edition
of the book, it remains, in an interesting way, problematic. By the
third edition, (1951), this problematic character has disappeared.[8]

What are the mechanisms of microevolution? In both editions
Dobzhansky lists three: (1) "mutations and chromosomal changes"[9];
(2) "the dynamic regularities of the physiology of populations,"
including "selection, migration, and geographical isolation"; and
(3) "the fixation of the diversity already attained on the preceding
two levels" (1937, p. 13; 1951, p. 18).[10] Strikingly, these mechanisms are
used to set aside the sorts of problems regarding macroevolution
that preoccupied Filipchenko. Here are the fundamental conclusions that Dobzhansky expressed on this topic in the final chapter of
the first edition (Dobzhansky, 1937): species are the only (and thus
the highest) systematic category to have withstood critical analysis
(p. 306); thanks to physiological isolating mechanisms such as the
chromosomal ones revealed by his own empirical studies, "discrete

[8] The sentence about the equality of mechanisms is replaced with the following
paragraph in the third edition: "Many authors believe that microevolutionary changes
are different in principle from macroevolutionary ones, and that while the former can
be understood in terms of the known genetic agents (mutation, selection, genetic
drift), the latter involve forces that are experimentally unknown or only dimly dis-
cerned. Views of this kind have been entertained by few geneticists (among whom
there is, however, so eminent a man as Goldschmidt, 1940), but they have been popular among those who approach evolutionary problems on the basis of data of paleontology and comparative anatomy. Well-known writers have supposed macroevolutionary changes to be engendered by some directing forces either inherent in the
organism itself or acting on it by some inscrutable means from outside. These guiding
forces received a variety of names, including orthogenesis, nomogenesis, aristogenesis, hologenesis, and finalism, but they escaped precise definition which would make
them subject to experimental test or to any kind of rigorous proof or disproof (see
Simpson 1949)" (Dobzhansky 1951, pp. 16–17).

[9] Both explicitly included in Philiptschenko (1927). One of the most important differences between Dobzhansky's and Filipchenko's treatments of microevolution is the
former's emphasis on chromosomal changes. The increased appreciation of chromosomal phenomena and their potential evolutionary significance is largely a consequence of the decade of empirical work that Dobzhansky had performed with various
colleagues, especially Sturtevant, in Morgan's laboratory. An examination of the technical papers of that decade or of the early papers in the GNP series is quickly convinc-
ing on this point. It should also be noted that the only sorts of variation that Dobzhansky investigates in his treatment of evolutionary statics in the first edition are genic
and chromosomal variation.

[10] The word *already* is omitted from the last quotation in the third edition.

groups of organisms frequently coexist in the same territory without losing their discreteness . . . [thus causing] a more or less permanent fixation of organic discontinuity" (p. 312); and "the conclusion that is inexorably forced on us is that the discontinuous variation encountered in nature, except that based on single gene differences, is maintained by means of preventing the random interbreeding of the representatives of now discrete groups" (p. 308). With these claims, apparently, the problems of macroevolution have been set aside.

From the First to the Third Edition

The above account sheds light on both the structure of Dobzhansky's 1937 book and its claims regarding macroevolution. But those claims were in many respects less controversial in the American context in the 1940s than others that played a central role in Dobzhansky's book. The shift in Dobzhansky's views is subtle, requiring careful examination of the details of his increasing selectionism and adaptationism. Let us examine some of the changes that occurred, especially those bearing on the importance and precise role of natural selection.

Of particular interest is the changing interplay between selection and "the scattering of variability," Dobzhansky's (1937) term for the effect of drift in small populations. In Chapter Five of the first edition, which contains the transition from evolutionary statics to evolutionary dynamics (and which was broken up and redistributed in the third edition), this topic is developed at length. There the scattering of variability is crucial; it ensures that differences between natural populations arise and are maintained in such a way as to allow selection to operate on those differences. Thus, in context, a claim made near the end of the chapter is quite important: "the available data, meager as they are, tend to show that the effective population sizes may prove to be small, at least in some species" (1937, p. 145).

Small size, as Sewall Wright had shown, allowed fixation of selectively neutral variation; it thus provided for the differentiation of populations. In the last section of the chapter, this Wrightian mechanism served as the basis for the formation of *microgeographic races* (a term coined by Dobzhansky, comparable to Filipchenko's *Jordanon*). Dobzhansky interprets such races as exhibiting nonadaptive differences that are probably not the product of natural selection (p. 148).

The predominant view in the 1937 edition and the first eight papers of GNP is that microgeographic races are expansions of small Wrightian demes, i.e., of populations with distinctive gene frequencies and constellations of traits produced mainly by isolation and genetic drift. GNP IX (1943), however, announced the discovery that the frequency of chromosome inversions in isolated populations of *Drosophila pseudoobscura* varies cyclically with the season. This article, together with a huge body of allied work pointing to larger population sizes and migration rates than Dobzhansky had anticipated, marks the beginning of the shift to the much more selectionist views expressed in subsequent articles in the GNP series and in the 1951 edition of *Genetics and the Origin of Species*. This change eliminates the need to rely on the scattering of variation (drift) to maintain microgeographical races. It thus marks the abandonment of an "internal dynamic" in the physiology of populations based on the intrinsic (if accidental) character of the populations over and above their interactions. Put hyperbolically, the "founder effect" (Mayr's term) produces variation whose fate in the hardened synthesis is solely in the hands of selection. More soberly, after the shift the "intrinsic character" of a population is mainly the product of selection.

This marks a major change from the problematic that Dobzhansky inherited from Filipchenko and which, as I have argued, helped structure the first edition of *Genetics and the Origin of Species*. Now the study of variation within and between groups is automatically a part of evolutionary *dynamics*, a part of the study of the changes wrought by selection and other evolutionary "forces." To this extent there is no reason for a separate study of evolutionary statics. Thus, the abandonment of the large-scale chapter organization separating statics from dynamics in the third edition makes good sense; the differences among the mechanisms at the different levels are not serious enough to bear much weight. The distinction between statics and dynamics has been subtly undermined by the reduced importance assigned to drift.

Although the point cannot be developed here, I suggest that this shift (together with many other factors) helped remove Dobzhansky's remaining ambivalence regarding the relation of micro- to macroevolution. Once he supposed that selection is dominant at all levels, he no longer had any ground for supposing that the intrinsic character of organisms *or populations* plays a significant role in macroevolution *independent of the scrutiny of selection.*

Concluding Questions

As I began, so let me end, by asking questions. Is it true, as I have suggested, that as late as 1937 Dobzhansky was still to some extent working out the consequences of abandoning views like those of Berg and Filipchenko in favor of those of Chetverikov, Serebrovskii, and Wright? Is it right that the reduced importance he came to assign to drift in maintaining differentiation of populations marks a more clear-cut allegiance to views like Serebrovskii's? Do these considerations, joined with those of Adams (1979, 1988, 1990f, and in preparation), help us understand the significance of Dobzhansky's introduction of the term *gene pool* into the language of evolutionary biology in 1950?[11]) And is the introduction of *gene pool* itself a marker of a phase shift in Dobzhansky's career, the one that Provine describes as a shift "away from [concern with] population structure, more toward the analysis of selection in nature and in the laboratory" (Provine 1986, p. 398)?[12] Although Dobzhansky himself would have resisted most of these suggestions (Dobzhansky 1980, p. 242), I think there is enough in them to be worth further investigation.

However one ultimately answers such questions, it is clear that we will not have a full understanding of Dobzhansky's work until we can integrate the Russian and American aspects of his career and thought. It is to be hoped that this volume will mark a fruitful step in that direction.

Acknowledgments

This paper has benefited greatly from discussions with participants in the Leningrad symposium on Theodosius Dobzhansky. Mark Adams and Daniel Alexandrov have both been of immense help in providing suggestions and criticizing errors. I am very grateful to them.

[11] As Adams argues, we must be careful to consider the full context for even narrowly scientific arguments: the pretension to have set aside the problems in the way of Darwinian extrapolationism clearly has an ideological component, both in the U.S. and in the USSR. See Adams 1988; and in preparation, pp. 35 ff.

[12] Although the third edition (1951) is far more selectionist than the first, there is still a very large difference between it and *Genetics of the Evolutionary Process* (Dobzhansky 1970) which is clearly a text in a new tradition, one that works within the hardened synthesis and tries to assimilate molecular data and techniques.

REFERENCES

Adams, Mark B. 1968. "The Founding of Population Genetics: Contributions of the Chetverikov School, 1924–1934." *Journal of the History of Biology* 1: 23–39.

―――. 1979. "From 'Gene Fund' to 'Gene Pool': On the Evolution of Evolutionary Language." In *Studies in History of Biology*, vol. 3, ed. W. Coleman and C. Limoges (Baltimore: The Johns Hopkins University Press), pp. 241–285.

―――. 1980. "Sergei Chetverikov, the Kol'tsov Institute, and the Evolutionary Synthesis." In E. Mayr and W. Provine, eds., *The Evolutionary Synthesis: Perspectives on the Unification of Biology* (Cambridge: Harvard University Press), pp. 242–278.

―――. 1988. "Rethinking the History of Population Genetics." Unpublished manuscript.

―――. 1990a. "Chetverikov, Sergei Sergeevich." In F. L. Holmes, ed., *Dictionary of Scientific Biography*, vol. 17, supp. 2 (New York: Charles Scribner's Sons), pp. 155–165.

―――. 1990b. "Eugenics in Russia, 1900–1940." In M. Adams, ed., *The Wellborn Science: Eugenics in Germany, France, Brazil, and Russia* (New York: Oxford University Press), pp. 153–216.

―――. 1990c. "Filipchenko [Philiptschenko], Iurii Aleksandrovich." In F. L. Holmes, ed., *Dictionary of Scientific Biography*, vol. 17, supp. 2 (New York: Charles Scribner's Sons), pp. 297–303.

―――. 1990d. "Karpechenko, Georgii Dmitrievich." In F. L. Holmes, ed., *Dictionary of Scientific Biography*, vol. 17, supp. 2 (New York: Charles Scribner's Sons), pp. 460–464.

―――. 1990e. "Serebrovskii, Aleksandr Sergeevich." In F. L. Holmes, ed. *Dictionary of Scientific Biography*, vol. 18, supp. 2 (New York: Charles Scribner's Sons), pp. 803–811.

―――. 1990f. "La génétique des populations était-elle une génétique évolutive?" In Jean-Louis Fischer and William H. Schneider, eds., *Histoire de la Génétique: Pratiques, Techniques et Théories* (Paris: A.R.P.E.M.), pp. 153–171.

―――. "Little Evolution, Big Evolution: Rethinking the Evolutionary Synthesis." Manuscript.

Dobzhansky, Theodosius. 1937. *Genetics and the Origin of Species*. New York: Columbia University Press.

―――. 1941. *Genetics and the Origin of Species*, 2d ed. New York: Columbia University Press.

―――. 1950. "Mendelian Populations and Their Evolution." *American Naturalist* 84: 401–418; reprinted in L. C. Dunn, ed., *Genetics in the 20th Century* (New York: Macmillan, 1951), pp. 575–589.

Dobzhansky, Theodosius. 1951. *Genetics and the Origin of Species* 3d ed. New York: Columbia University Press.

——. 1970. *Genetics of the Evolutionary Process* (New York: Columbia University Press), pp. 71–93.

——. 1980. "The Birth of the Genetic Theory of Evolution in the Soviet Union in the 1920s." In E. Mayr and W. Provine, eds., *The Evolutionary Synthesis: Perspectives on the Unification of Biology* (Cambridge: Harvard University Press), pp. 229–242.

Filipchenko, Iurii A. 1929. *Izmenchivost' i metody ee izucheniia* [Variation and methods for its study], 4th ed. Leningrad: Gosizdat.

Gaissinovitch, A. E. 1980. "The Origins of Soviet Genetics and the Struggle With Lamarckism, 1922–29." Trans. Mark B. Adams. *Journal of the History of Biology* 13: 1–51.

Goldschmidt, Richard B. 1940. *The Material Basis of Evolution.* New Haven: Yale University Press.

Gould, Stephen Jay. 1982. "Introduction." In *Genetics and the Origin of Species* (the reissued text of Dobzhansky 1937) (New York: Columbia University Press), pp. xvii–xxxix.

——. 1983. "The Hardening of the Evolutionary Synthesis." In M. Grene, ed., *Dimensions of Darwinism.* New York: Cambridge University Press.

Lewontin, R. C., J. A. Moore, W. B. Provine, and B. Wallace, eds. 1981. *Dobzhansky's Genetics of Natural Populations I–XLIII.* New York: Columbia University Press.

Philiptschenko, Jurii. 1927. *Variabilität und Variation.* Berlin: Gebrüder Borntraeger.

Provine, William B. 1986. *Sewall Wright and Evolutionary Biology.* Chicago: University of Chicago Press.

Serebrovskii, A. S. 1926. "Teoriia nasledstvennosti Morgana i Mendelia i marksisty" [Marxists and the theory of heredity of Morgan and Mendel]. *Pod znamenem marksizma* 3: 98–117.

——. 1928. "Genogeografia i genofond sel'skokhoziaistvennykh zhivotnykh SSSR" [Genogeography and the genofund of agricultural animals of the USSR]. *Nauchnoe slovo,* no. 9: 3–22.

Todes, Daniel P. 1989. *Darwin Without Malthus: The Struggle for Existence in Russian Evolutionary Thought.* New York: Oxford University Press.

PART THREE
THE SCIENTIFIC LEGACY

■

Dobzhansky, Waddington, and Schmalhausen:
Embryology and the Modern Synthesis

Scott F. Gilbert

I

This paper seeks to outline the attempts by Ivan Ivanovich Schmalhausen and Conrad Hal Waddington to integrate embryology with
the Modern Synthesis and to understand why Dobzhansky favored
and popularized Schmalhausen's work but denigrated Waddington's. The fact that Dobzhansky did approve of one over the other
suggests that although Waddington's and Schmalhausen's respective
syntheses were very similar, the differences between them were critical enough for Dobzhansky to perform some selection on them; the
differences were not neutral. What are these differences, and why
were they important to Dobzhansky?

II

In the 1930s genetics was rapidly replacing embryology as the premier way of studying heredity. The two fields were formally separated in T. H. Morgan's book *The Theory of the Gene* (1926): the geneticists were to study the transmission of hereditary traits, whereas
the embryologists were to study their expression. From 1915 to 1932,
new university chairs in genetics were created, new genetics journals
were established, and new genetics societies were founded. By 1938
the two fields had their own journals, their own professors, their own
vocabularies, their own rules of evidence, their own favored organisms, their own textbooks, and their own paradigmatic experiments.
Genetics books no longer mentioned development, and embryology
texts no longer mentioned genes.[1]

[1] See Jan Sapp, *Beyond the Gene: Inheritance and the Struggle for Authority in Genetics* (New York: Oxford University Press, 1987); and essays in *The American Development of Biology*, ed. R. Rainger, K. R. Benson, and J. Maienschein (Philadelphia: University of Pennsylvania Press, 1988), notably S. F. Gilbert, "Cellular Politics: Ernest

The Modern Synthesis is "modern" because it supplanted older syntheses. These older syntheses were also attempts to integrate heredity and evolution, but they forged evolution with embryology. The most famous of these older syntheses was the Biogenetic Law of Ernst Haeckel, who saw a very close relationship between ontogeny and phylogeny. Development took an embryo only so far. By adding a new step, however, development would occasion the production of a new organism. Because of this, ontogeny recapitulated phylogeny. In Russia this relatively crude synthesis was given new sophistication by Eli Metchnikoff, who was able to integrate it with A. Kovalevsky's observations of the formation of endodermal organs, Darwin's notion of branching phylogenies, and von Baer's law that animal development progressed from general forms to individualized forms.[2] Metchnikoff's framework still serves as the foundation of phylogenetic studies and is the basis of Libby Hyman's and Leo Buss's phylogenetic hypotheses, but it provided no internal mechanics of development. Haeckel's developmental mechanisms, like Weismann's before him, were found to be wanting. As Garstang and De Beer pointed out, ontogeny rarely ever could account for terminal additions of successive developmental stages. The traditional synthesis of embryology and evolution fell apart just as the syntheses between population genetics and evolution were beginning.

It has frequently been stated that embryology was left out of the Modern Synthesis. Not only have modern critics such as Gould, Wake, Adams, and Provine pointed this out, but so had Dobzhansky, Schmalhausen, and Waddington. In his forward to Schmalhausen's *Factors of Evolution*, which he arranged to have published in the United States, Dobzhansky wrote: "The book of I. I. Schmalhausen advances the synthetic treatment of evolution starting from a broad base of comparative embryology, comparative anatomy, and the mechanics of development. It supplies, as it were, an important missing link in the modern view of evolution." Dobzhansky also noted—and this is important for discerning his view of the relationship between embryology and genetics—that Schmalhausen's "command of genetics permits him to give a penetrating as well as inclusive analysis

Everett Just, Richard B. Goldschmidt, and the Attempt to Reconcile Embryology and Genetics" (pp. 311–346) and D. Paul and B. Kimmelman, "Mendel in America: Theory and Practice, 1900–1919" (pp. 281–310)

[2] A. I. Tauber and L. Chernyak, *Metchnikoff and the Origins of Immunology: From Metaphor to Theory* (New York: Oxford University Press, 1991).

of the developmental relationship in terms of genetic causation."[3] This had been the geneticists' credo in separating from embryology: Only the geneticist truly understands causal embryology.

Schmalhausen also saw that the Modern Synthesis had excluded functional morphology and embryology, and he stated this in no uncertain terms. After mentioning the work of Lloyd Morgan and Baldwin, he lamented, "Unfortunately, the more recent advances of genetics have promoted the spread of neo-Darwinian concepts."[4] While this might seem a criticism of Dobzhansky's 1937 definition of evolution as "a change in the genetic composition of populations,"[5] actually it was only a lament that the population genetics model of evolution failed to include material from embryology that would strengthen it. Schmalhausen was not saying that the Modern Synthesis could not explain evolution.

Other scientists, however, *were* saying that the Modern Synthesis, by itself, could not explain evolution—that embryology was needed not merely to strengthen the population genetics model of changes in gene frequency over time but rather to complement the models of the population geneticists. Population genetics was fine for microevolution, but it could not account for macroevolution. It could account for the survival of species but not for the arrival of species. It could explain how a peppered moth became a dark gray moth but not how an annelid could become an arthropod.[6]

The most vocal critic of the neo-Darwinian synthesis was Richard Goldschmidt, who in 1940 exclaimed: "I may challenge the adherents of the strictly Darwinian view, which we are discussing here, to try to explain the evolution of the following features by the accumulation and selection of small mutations: hair in vertebrates, feathers in birds, segmentation in arthropods and annelids, the transformation of the gill arches in phylogeny, including the aortic arches, muscles, nerves, etc. further, teeth, shells of mollusks, ectoskeletons, com-

[3] Th. Dobzhansky, "Foreword," in I. I. Schmalhausen, *Factors of Evolution: The Theory of Stabilizing Selection*, trans. I. Dordick, ed. Th. Dobzhansky (1949; reprint Chicago: University of Chicago Press, 1986).

[4] Schmalhausen, *Factors of Evolution* (1986 reprint), p. 198.

[5] Th. Dobzhansky, *Genetics and the Origin of Species* (New York: Columbia University Press, 1937), p. 11.

[6] See Mark B. Adams, "La génétique des populations était-elle une génétique évolutive?" in *Histoire de la Génétique: Pratiques, Techniques et Théories*, ed. Jean-Louis Fischer and William H. Schneider (Paris: A.R.P.E.M and Éditions Sciences en situation, 1990), pp. 153–171; and Mark B. Adams, "Little Evolution, Big Evolution: Rethinking the Evolutionary Synthesis," in preparation.

pound eyes. . . ." And he went on. However, Goldschmidt's own synthesis of evolution, genetics, and development found few adherents. It left out most of modern genetics since Goldschmidt did not believe the gene to be a discrete entity.

On a less strident scale, C. H. Waddington also sought to complement the population genetics view with that of embryology. He would frame a hypothesis called *genetic assimilation* to account for certain macroevolutionary changes. This hypothesis would "provide a plausible account of the result in terms of orthodox genetic and embryological mechanisms."[8] The orthodox genetics was not sufficient to do so alone.

III

The syntheses of embryology, genetics, and evolution put forth by Waddington in 1940, 1941, and 1953 and by Schmalhausen in 1949 are usually linked together. Viktor Hamburger said that "Waddington in England had independently developed ideas similar to those of Schmalhausen during the war."[9] Like Dobzhansky, Waddington also used the "missing link" trope to describe the absence of embryology from the Modern Synthesis. (Interestingly, this trope is Haeckelian; only if there is a linear chain can there be a missing link.) Similar statements linking the syntheses of Waddington and Schmalhausen can be seen throughout the literature on the history of the Modern Synthesis.[10]

Both Schmalhausen and Waddington had received excellent, broad training. Waddington (1905–1975) started his degree in paleontology, working on ammonites and nautiloids. He then went into embryology and trained with Spemann in Freiburg. In 1937 Waddington went to Caltech where he worked under Sturtevant and wrote one of the first papers in *Drosophila* developmental genetics—on the

R B Goldschmidt *The Material Basis of Evolution* New Haven Yale University Press 1940 pp 6–7

C H Waddington "Canalization of Development and the Inheritance of Acquired Characteristics" *Nature* 150 1942 : 563–565.

V Hamburger "Embryology and the Modern Synthesis in Evolutionary Theory," in *The Evolutionary Synthesis. Perspectives on the Unification of Biology*, ed. E Mayr and W Provine Cambridge: Harvard University Press 1980 pp 97–112.

E g W B Provine "Embryology" in *The Evolutionary Synthesis* ed Mayr and Provine pp 96–97 and D B Wake "Foreword 1986" in Schmalhausen. *Factors of Evolution* 1986 reprint pp. v–xii; see Note 3 genes controlling wing development—and the textbook *Introduction to Modern Genetics* (London: Allen and Unwin, 1939). He felt that evolution, embryology, and genetics were all one discipline, which he called *diachronic biology*.[11]

Schmalhausen (1894–1963) was a student of A. N. Severtsov, the famed evolutionary morphologist whom Schmalhausen succeeded as head of the Institute of Evolutionary Morphology. Severtsov knew genetics well enough to use its evidence against the Lysenkoists. However, he did not think that population genetics explained macroevolution. In a work probably written in 1936 but published only posthumously in 1939, Severtsov wrote: "Despite the brilliant successes in hereditary theory, the results of genetics research have brought little to the solution of evolutionary problems. Experimental embryologists have done even less in this direction."[12] Schmalhausen was to remedy this deficiency. In addition to his research on the origin and embryology of vertebrates, he wrote a leading textbook on Darwinism and became familiar with the population genetics that was playing such a large role in Soviet evolutionary thinking. Even in his embryology book, *The Organism as a Whole in Individual and Historical Development*, Schmalhausen cited Dubinin, Goldschmidt, Balkashina, Sumner, Wright, and Timoféeff-Ressovsky.[13] It was because he knew so much genetics that the Darwinian embryologist Schmalhausen ran afoul of the Lysenkoists.

Both Schmalhausen and Waddington, then, were well prepared to attempt a synthesis of evolution, genetics, and embryology, and there are several reasons to link the theories of these two scientists together.

First, they both took embryology seriously and felt that the current version of the Modern Synthesis was incomplete without it.

Second, they both presented dialectical models. They stressed the interactions between groups of cells in forming organs, the interac-

[11] On Waddington's development, see C. H. Waddington, "The Practical Consequences of Metaphysical Beliefs on a Biologist's Work: An Autobiographical Note," in *The Evolution of an Evolutionist* (Ithaca, N.Y.: Cornell University Press, 1975), pp. 3–9; A. Robertson, "Conrad Hal Waddington: 8 November 1905–26 September 1975," *Biographical Memoirs of Fellows of the Royal Society* 23: 575–622; and S. F. Gilbert, "Induction and the Origins of Developmental Genetics," in *A Conceptual History of Modern Embryology*, ed. S. F. Gilbert (New York: Plenum Press), pp. 181–206.

[12] M. B. Adams, "Severtsov and Schmalhausen: Russian Morphology and the Evolutionary Synthesis," in *The Evolutionary Synthesis*, ed. Mayr and Provine, pp. 193–225.

[13] Ibid., p. 219.

tions between heredity and environment in producing the pheno-type (and in Waddington's case, the interactions between nucleus and cytoplasm in producing the cell type).

Third, they both used the language of systems in their analysis. Both would, at the end of their respective careers, become interested in cybernetics. Unlike other authors, these two had a formalism set up by their dialectic. It didn't really matter whether one was talking about cells, molecules, or organisms—they were all treated as systems that interact.

Fourth, they both hypothesized a model for the channeling of possible traits into a relatively narrow allowable range. Waddington called this *canalization*; Schmalhausen called it *stabilization*.[14] In 1941 Waddington wrote that "developmental reactions . . . are in general canalized. That is to say, they are adjusted so as to bring about one end regardless of minor variations in conditions during the course of the reaction."[15] Canalization can be considered a buffering of the developmental pathway by the interaction of several genes such that none of them is of crucial importance. The effect would limit the variations in a population such that "If wild animals of almost any species are collected, they will usually be found to be 'as like as peas in a pod'."[16] Once development had found a successful way of constructing an organism, it evolved so as to produce that phenotype with the least possible susceptibility to change.

This canalization became the hallmark of Schmalhausen's hypothesis for stabilizing selection. Indeed, it is ironic that in one part of Schmalhausen's book, his English translator (Isadore Dordick) used Waddington's term to depict this phenomenon: "Such reactions are canalized into the narrow channel of a more specific norm which is adapted to definite conditions of the external environment. . . ."[17] But in general, where Waddington would have used the word *canalized*, Schmalhausen would use the word *stabilized*.

Schmalhausen identified several mechanisms for this stabilization.[18] The first two are genetic—(1) diploidy, which provides stability in that an organism has two chances of having at least one functional allele for a certain product; and (2) the complexity of the

[14] Schmalhausen, *Factors of Evolution* (1986 reprint; see Note 3), p. 205.
[15] Waddington, "Canalization" (1942).
[16] Ibid.
[17] Schmalhausen, *Factors of Evolution* (1986 reprint), p. 195.
[18] Ibid., pp. 230–231

balanced genetic system, wherein variations can occur without lethal consequences. The remaining seven mechanisms involve embryological considerations and can be grouped into four arguments:
(A) There appears to be a reserve supply of morphogenic substances;
a little more or less does not make a difference to the organism.
(B) The role of individual morphogenic factors is small, since there
can be multiple factors, each adding to the concentration needed to
produce the developmental reaction. (C) There is a transformation
of specific morphogens into nonspecific activators. (This point was
emphasized by both Schmalhausen and Waddington and will be discussed below.) (D) In many embryos, especially among vertebrates,
the regulative capacity of the organism and the egg as a whole en-
ables the embryo to form an entire organism, even when a part is
missing.

The fifth similarity between Schmalhausen's and Waddington's
hypotheses was that they maintained that physiological adaptations
can be taken over by the genome. For Schmalhausen, this was another example of stabilizing selection. He used the concept of stabilizing selection as a framework to discuss the assimilation of physiological adaptations into embryonic adaptations. Schmalhausen
argued that organisms can respond physiologically to changes in
their environment. Animals that carry loads get larger muscles; animals that abrade the ground get calluses. These are developmental
changes, but they are produced in response to an outside stimulus.
As might be expected, individuals within a species would differ in
having muscles and calluses. After all, Arnold Schwarzenegger and I
are of the same species. This variation provides the first mechanism
for genetic stabilization—since, if selection pressure exists for one
end of the variation, a new norm would appear. Those at one end of
the original norm would survive, those at the other end would not.

The second mechanism for genetic stabilization would be to take
the ability to form muscles or calluses from the environmental inducer and transfer that ability to an embryological inducer. My descendants would be born more muscular because my ability to grow
muscles when I exercise would be transferred to the ability to respond to a chemical in the embryo. Schmalhausen said: "Those
physiological and functional adaptations, which acquire permanent
value under certain conditions of existence, are incorporated into
the organization, and become permanent as a result of the continuous action of stabilizing selection. At the same time, the system

of morphogenetic correlations increases in complexity as the internal factors of individual development are transformed from dependent differentiation into more autonomous self-differentiating systems."[19]

What Schmalhausen saw is the transformation from an environmentally dependent differentiation which is an accident of conditions peculiar to an individual organism to a normative part of embryonic development characteristic of the species. Individual variation has diminished as the trait becomes genetically stabilized into the population.

> Thus in the course of evolution, the organism is emancipated from harmful types of dependence upon external factors and both its vital processes and its ontogeny become more and more autonomous.... Functional factors [physiological, environmentally induced ones] have been replaced by hormonal factors which in turn have given way to morphogenetic ones. Local influences, which had been merely accessories to the development of functional differentiations have become indispensable conditions of development.... All these changes have been the result of stabilizing natural selection. Hence, stabilizing selection is the most important agent altering the factors of individual development, determining the continuous process whereby individual adaptations are gradually incorporated into normal organization, and consequently, transforming all of ontogeny by progressively raising the regularity of normal morphogenesis and the stability of the adapted norm.[20]

Waddington had been saying much the same thing. He called it (most unfortunately) *genetic assimilation*. Early in Waddington's career, he, Needham, and Brachet made a disturbing discovery. They found that not only could compounds from other parts of the body mimic the natural embryonic inducer of neural tissue in amphibians, so too could many artificial compounds that had no relationship to any known natural chemical. This lack of specificity halted their attempts to discover the natural inducer molecule.[21]

However, Waddington learned a lesson from this. What was critical was not the inducer molecule—several disparate chemicals could

---

[19] Ibid., p. 234.

[20] Ibid , pp. 238, 242.

[21] C. H. Waddington, J. Needham, and J. Brachet, "Studies on the Nature of Amphibian Organisation Center: III. The Activation of the Evocator," *Proceedings of the Royal Society of London (Biology)* 120: 173–198 For further discussion of this, see also Gilbert, "Cellular Politics" (see Note 1).

induce the ectoderm to form a neural tube. Rather what was critical was the competence of the responding cell to be induced. Not every cell could become a neuron when exposed to natural or artificial inducers. This led him to the notion of canalization, which he used to define those developmental pathways inherent within the competent cells. It also led him to the notion of genetic assimilation, which he first published in 1941. If the ability to respond is what is critical, and if several agents can induce this response, then the important part of evolution is to evolve *the ability to respond.* For instance, if the ostrich forms calluses where its feet abrade the ground, the ostrich skin is said to have evolved the competence to form calluses. This ability to respond to friction is assumed to be a selectable trait. Those ostriches that can form calluses have a better chance of surviving than those that do not. The next step would be to transfer this competence from being induced by friction to being induced by some embryological agent. Since Waddington felt that the inducer could be relatively nonspecific, this would not be difficult. The important thing was to get that competence in the responding tissue.

As Waddington wrote, "Presumably its skin, like that of other animals, would react directly to external pressure and rubbing by becoming thicker. . . . This capacity to react must itself be dependent upon genes. . . . It may then not be too difficult for a gene mutation to occur which will modify some other nearby area in the embryo in such a way that it takes over the function of external pressure, interacting with the skin so as to 'pull the trigger' and set off the development of the callousities."[22] Thus, Waddington felt that these and other examples of Lamarckian "inheritance of acquired characteristics" could be explained by orthodox natural selection and by orthodox embryology.

Schmalhausen also used embryological evidence (although not Waddington's) to show that induction is not specific. Moreover, Schmalhausen also used the example of the ostrich calluses and to the same effect: "Clearly, callosities which first arose as modifications in response to a local stimulus have subsequently begun to develop at the same locus in the absence of the external stimulus. Thus, a new structure which was first produced by the direct differentiation of a functional stimulus has become stabilized."[23]

Both men thought that physiological induction could be trans-

[22] Waddington, "Canalization" (1942)
[23] Schmalhausen, *Factors of Evolution* (1986 reprint), p. 204.

ferred to embryological induction. To Waddington this was an example of genetic assimilation; to Schmalhausen it was an example of genetic stabilization.

IV

These five important similarities—dialectics, systems approach, genetic stabilization, genetic assimilation, and the desire to reincorporate embryology into the studies of evolutionary mechanism—certainly set Waddington and Schmalhausen off from the rest of the biological theorists of their time. What, then, would cause Dobzhansky to favor one over the other? While he praised Schmalhausen's book and recommended its translation and printing in the United States, he downplayed Waddington's similar studies.

Waddington was quite annoyed at Dobzhansky's repeated citation of Schmalhausen's work over his own, and he wrote so to Dobzhansky. In July 1959 he complained to Dobzhansky that "whenever you refer to someone on developmental buffering and such subjects, you always quote Schmalhausen and never my work." Waddington acknowledged that his theories and those of Schmalhausen were very similar and were probably formed independently and at about the same time, but all the same, Waddington claimed, his own theory was based on more solid data—his "detailed study of 40 genes affecting the *Drosophila* wing, in particular as discussed in my *Organisers and Genes* of that year—while in his [Schmalhausen's] case it was more of a matter of theory to bring genes into the story."[24] Dobzhansky replied the following month: "Having thought about it, I plead guilty and apologize." However, after saying this he defended his earlier decision to play up Schmalhausen's work. He told Waddington that he regarded *Factors of Evolution* as one of the "basic books" establishing the biological theory of evolution and that, as a victim of Lysenko, Schmalhausen needs "any support that we may give him."[25]

If Dobzhansky apologized in private to Waddington, his public statements still left no doubt that he favored Schmalhausen's theories over Waddington's. In Dobzhansky's *Genetics of the Evolution-*

[24] Letter from C. H. Waddington to Theodosius Dobzhansky, 25 July 1959; published in *The Evolution of an Evolutionist* (Ithaca, N.Y.: Cornell University Press, 1975), pp. 96–97

[25] Letter from Theodosius Dobzhansky to C. H. Waddington, 15 August 1959; published in *The Evolution of an Evolutionist* (1975), pp. 96–97.

ary Process (1970), Waddington is mentioned favorably, along with Schmalhausen, in connection with canalization and genetic stabilization.[26] However, when it came to the assimilation or stabilization of physiologically induced traits, Dobzhansky presented Waddington as a frustrated Lamarckian. He summarized an experiment that Waddington had published in 1953 where he claimed to have shown the genetic assimilation of a wing vein trait in *Drosophila* in the laboratory. Dobzhansky believed that this evidence of "so called 'genetic assimilation'" was really due to the selection of preexisting variants from the original population. He concluded: "The analogy with alleged Lamarckian inheritance is superficial, at best, as Waddington clearly realizes."[27] In this context, Schmalhausen escapes unmentioned.

But as we have seen, especially in the 1940s, Waddington could not be considered as having Lamarckian leanings, and such genetic assimilation was also part of Schmalhausen's theory of genetic stabilization. Waddington had gone into the laboratory to try to prove that genetic assimilation could be shown in *Drosophila* cages. Dobzhansky quoted Waddington's experimental work but not the theoretical foundations of the work published in 1941.

I believe that the difference between the hypotheses of Waddington and Schmalhausen is that Waddington presented his theory as a complement to population genetics. Population genetics alone was not sufficient to explain evolution. The definition of evolution had to be broadened to include more than just changes in gene frequency. Schmalhausen, however, was insistent that genetic stabilization should be made part of the population genetical model evolution. For him, embryological evidence did not fall outside the current definition of evolution; it fell nicely within it. Unlike Waddington, who stressed that genetic assimilation could give new types of organisms, Schmalhausen emphasized that "In these instances of adaptation, nothing new actually arises."[28]

Waddington saw embryology as complementing the Modern Synthesis; Schmalhausen saw embryology as completing it. It is no wonder, then, that Dobzhansky would find Schmalhausen's studies more suitable and better able to be integrated into the Modern Synthesis. Schmalhausen's book provided Dobzhansky with his "missing

---

[26] Th. Dobzhansky, *Genetics of the Evolutionary Process* (New York: Columbia University Press, 1970), pp. 40, 96

[27] Ibid., p. 211.

[28] Schmalhausen, *Factors of Evolution* (1986 reprint), p. 200.

link."[29] Embryology fit well into the synthesis. Waddington's work was no missing link; it claimed that another chain had to be constructed. Therefore, in addition to personal and moral reasons, there were also substantive reasons for Dobzhansky's preference of one synthesis over another.

This type of observation has been seen before. Indeed, Fred Churchill, writing about another pair of biologists who worked on the synthesis of evolution and embryology, came to the same conclusion. These biologists were Gavin De Beer and Julian Huxley. De-Beer, noted Churchill, approached the synthesis as an embryologist and saw an importance for ontogeny in forming new organisms. Huxley, however, "quickly saw that population genetics and the study of natural populations filled the very same causal role vacated by the much discredited biogenetic law."[30]

As contemporary biologists attempt to unravel the relationships between genetics, development, and evolution, this typology still exists. There are those biologists who see the genetics of microevolution as sufficient for explaining the morphological changes of macroevolution. Whatever changes occur to make evolution possible would have to occur by the known molecular genetic mechanisms. Other scientists claim that the evolutionary theory is incomplete without a knowledge of those developmental principles that could explain the origins of dramatically new phenotypes, and that these principles would not necessarily be those of molecular or population genetics. Dobzhansky realized that Waddington's theory and Schmalhausen's theory, although extremely similar, separated on this major point.

ACKNOWLEDGMENTS

I would like to express my special thanks to the organizers of the Leningrad symposium on Dobzhansky. I would also like to acknowledge my debt to Mark Adams and to his writings on population genetics, Dobzhansky, Schmalhausen, and the microevolution/macroevolution problem, on which I have heavily relied.

[29] On Dobzhansky and Schmalhausen, see Mark B. Adams, "More Than Dialectics . . . ," *Isis* 79, 297 (June 1988): 281–284.

[30] F. B. Churchill, "The Modern Evolutionary Synthesis and the Biogenetic Law," in *The Evolutionary Synthesis*, ed. Mayr and Provine (see Note 9), pp. 112–122.

■

Theodosius Dobzhansky Remembered:
Genetic Coadaptation

Bruce Wallace

I WAS A last-year undergraduate student at Columbia University when I met Dobzhansky. He had inquired of Arthur Pollister whether any Columbia student would be interested in employment collecting *Drosophila pseudoobscura* during the summer of 1941. John A. Moore, when asked by Pollister, recalled that as a freshman I had remarked that I was majoring in zoology in order to travel. Before introducing me to Dobzhansky, however, Pollister gave me a copy of *Genetics and the Origin of Species* (Dobzhansky 1937) and insisted that I read it. Only then would Pollister make the introduction.

My collaboration, or at least correspondence, with Dobzhansky extended from 1941 until his death in 1975. Two summers were spent collecting flies in the San Jacinto region of southern California, dissertation work was carried out in his laboratory after World War II, and then came an extended interaction with him, his students, and foreign associates during the eleven years I spent at Cold Spring Harbor, only one hour by train from New York City. After my move to Cornell University in 1958, my contact with the Dobzhanskys became less frequent; except for letters, it virtually ceased after their move to California.

Dobzhansky's love of his motherland and of Russian culture never slackened during all the years I knew him. The evil events of 1942 depressed him terribly and, except for valiant efforts on the part of his wife, Natasha, and myself, the summer's research on the dispersion of flies would have been abandoned because of his despair. He was extremely disappointed at not being allowed to participate in the scholarly exchange that was arranged by the two academies of science. And as the end of his life approached, he lamented to I. Michael Lerner that with no one except Lerner could he communicate in Russian; apparently he could properly express his feelings about his approaching death only in his mother tongue.

Although the concept of the coadaptation of organisms to each other's presence or of parts of individual organisms to other parts was mentioned explicitly by Darwin (1859), Dobzhansky (1950) seems to have first introduced the concept into experimental population genetics.

The immediate start of Dobzhansky's interest in coadaptation can be ascribed to his studies of the relative fitnesses of individuals homozygous or heterozygous for certain naturally occurring inversions in *Drosophila pseudoobscura* (Wright and Dobzhansky 1946). Before the publication of these experimental results, Dobzhansky tended to view different gene arrangements as being selectively neutral (Dobzhansky and Queal 1938; Dobzhansky and Epling 1944). Contrary to earlier indications, carriers of different gene arrangements proved to exhibit large differences in overall fitness, at least in some environments. Typical results of these early experiments are shown in Table 1.

The early results revealed that when differences in fitness were observed, inversion heterozygotes were generally superior. They also revealed that the relative fitnesses of corresponding genotypes (where the chromosomes were obtained from different geographic localities) were not always the same. Thus, in populations containing the gene arrangements ST and CH, the relative fitnesses of ST/ST individuals varied from 0.78 to 0.91 and CH/CH individuals from 0.28 to 0.58 depending upon the source of the chromosomes (Piñon Flats, Keen Camp, and Mather—all localities in California). This sort of variation was regarded as proof that the gene content of chromosomes with the same gene arrangement varied, thus ruling out position effect as the only factor governing the retention of a particular gene arrangement in (or its elimination from) a population.

To account for the pervasive superiority of inversion heterozygotes, knowing that the gene content of chromosomes with the same gene arrangement can vary, Dobzhansky (1950) was led to a study of laboratory populations in which the chromosomes having different gene arrangements came from different geographic localities. The results of several of these experiments are shown in Table 2. The contrast with Table 1 is immediately apparent. In the geographically mixed populations, heterozygotes were nearly always intermediate in fitness; in the single exception, these flies ostensibly had the lowest of the three fitnesses.

TABLE 1

Relative Fitnesses of a Number of Homozygous and
Heterozygous Genotypes Involving Gene Arrangements
of *D. pseudoobscura* Obtained from a Number of
Localities within California

Locality	Genotype				
	1	2	1/1	1/2	2/2
Piñon	ST	CH	0.85	1.00	0.58
Keen	ST	CH	0.91	1.00	0.42
Mather	ST	CH	0.78	1.00	0.28
Piñon	ST	AR	0.81	1.00	0.50
Keen	ST	AR	0.79	1.00	0.58
Mather	ST	AR	0.64	1.00	0.58
Piñon	AR	CH	0.86	1.00	0.48
Keen	AR	CH	0.54	1.00	0.60
Mather	AR	CH	0.81	1.00	0.60
Mather	AR	TL	0.69	1.00	0.12
Mather	ST	TL	1.12	1.00	0.33

TABLE 2

Relative Fitnesses of Individuals Homozygous and Heterozygous
for Various Gene Arrangements in Laboratory Populations of
*D. pseudoobscura**

Localities	Arrangement		Fitnesses		
	1	2	1/1	1/2	2/2
Mather versus Piñon					
45	AR^M	CH^P	1.28	1.00	0.47
46	ST^P	AR^M	0.63	1.00	1.51
47	ST^M	AR^P	1.31	1.00	0.57
48	ST^M	CH^P	1.18	1.00	0.48
49	ST^P	CH^M	1.38	1.00	0.43
Mexico versus Piñon					
55	ST^P	CH^{Mex}	1.26	1.00	0.87
56	AR^P	CH^{Mex}	1.53	1.00	1.16

* Unlike those listed in Table 1, these are geographically mixed populations; the superscript identifies the geographic origin of each gene arrangement as Mather, Piñon, or Mexico. (Dobzhansky, 1950.)

The experimental procedure by which the data of Table 2 were obtained should be noted. These fitnesses were estimated from the deviations of observed frequencies of 1/1, 1/2 and 2/2 flies in the F_2 generation from the 1:2:1 ratio expected following the mating of 1/2 × 1/2 F_1 hybrids. The tested larvae developed, of course, in grossly overcrowded food cups of the sort used in population cages; the numerous eggs on these cups were laid by the females among 1000–2000 optimally reared, parental hybrid flies.

The above account is emphasized because many persons believe that Dobzhansky demonstrated that interpopulation inversion heterozygotes are inferior to interpopulation inversion homozygotes even among F_1 interpopulation hybrids. No evidence bearing on this question is to be found among Dobzhansky's experiments: F_1 interpopulation hybrids heterozygous for different gene arrangements were obtained under circumstances that did not permit measurements of relative viabilities; these flies were merely used as parents in order to obtain a segregating generation. Deviations from Mendelian ratios allowed Dobzhansky to estimate relative fitnesses. Inversion heterozygotes proved to be intermediate in fitness on a mixed-population genetic background.

The estimates of relative fitnesses presented in Table 2 are based solely on egg-to-adult survival (plus a small portion that might have reflected a differential success of adult flies when mated with the known AR/AR tester strain). This is obviously an important component of total fitness but not necessarily the only component. Therefore, Dobzhansky (1950) started experimental populations with mixed locality F_1 hybrid flies that were heterozygous for differing gene arrangements.

The results obtained by studying these populations were highly variable. In some, one gene arrangement displaced the other; others established equilibria but at times only after undergoing considerable fluctuations in inversion frequency over time. Such results led to further studies, such as one with Howard Levene (Dobzhansky and Levene 1951) entitled "Development of heterosis through natural selection in experimental populations of *Drosophila pseudoobscura*" and another (Dobzhansky and Pavlovsky 1953) entitled "Indeterminate outcome of certain experiments on *Drosophila* populations."

Genetic coadaptation in Dobzhansky's sense, then, consisted of those changes that resulted in and virtually assured the perpetuation of the superiority of inversion heterozygotes. Chromosomal poly-

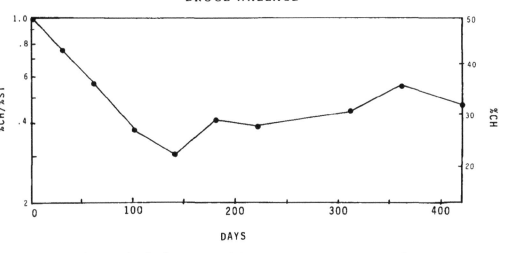

Fig. 1. Changes in the frequency of the CH gene arrangement (right) and %CH/%ST ratio (left) in a mixed-locality experimental population. Chromosomes of the CH gene arrangement were obtained from Mexico, those with the ST gene arrangement from California. (After Dobzhansky and Levene 1951.)

morphism was an essential part of his coadapted gene pool. I do not recall any statement to the effect that, in those instances in which one gene arrangement displaced another, coadaptation may have occurred. This paper is concerned with Dobzhansky's views on co-adaptation, not my own. Still, I might take a moment to mention that through our concurrent studies of irradiated populations of *Drosophila melanogaster* (and later of DDT resistance in this same species), J. C. King and I came to view the entire gene pool of an isolated population as one in which events occurring at one gene locus had unavoidable consequences affecting the selection for or against alleles at other loci. Hence, coadaptation is a term that, in our view, can be applied to the entire gene pool: the integration of gene pools, in a general sense. Because we inferred that many of the events we witnessed in irradiated populations involved selection favoring heterozygous individuals, I visualized much of this integration as occurring with respect to "heterotic" alleles—alleles exhibiting what I later called "marginal overdominance."

Although Dobzhansky may have had certain reservations about this view of populations at the outset ("I would rather believe Muller" is a comment I recall his making during one discussion), he did recognize the importance of genetic background with respect to coadaptation. In his population #66, the frequency of CH fell from

50 percent to less than 25 percent only to rise once more. Thus, selective forces resulted in the frequency of CH twice passing through 30 percent, for example: first decreasing, then increasing. A double passage of this sort could only happen in response to changing genetic backgrounds (see Lewontin 1974, p. 274).

Relatively late in his career (Dobzhansky and Spassky 1968), Dobzhansky emphasized the importance genetic background might play in determining the viability of heterozygotes for "recessive" lethals. By a series of laborious back crosses, he and Spassky created tester stocks whose genetic backgrounds were either of California or Arizona origins. They then tested lethals from these two regions for their effects on heterozygous carriers either within their normal background or the "wrong" background. Their results (confirmed as well by Anderson 1969) showed that, on the average, lethals in their native backgrounds enhanced the viability of their heterozygous carriers, whereas they had somewhat deleterious effects when tested on a foreign background. Although the phrases "epistatic interactions" and ". . . lethal chromosomes interact with the genetic background . . ." appear in the Dobzhansky-Spassky and Anderson publications, the term "coadaptation" does not.

In a book (Lewontin et al. 1981) of which I am one of four coeditors, Lewontin recalls that "the theory of coadaptation pretty much disappeared from Dobzhansky's work after 1955." The serious blow to the theory of coadaptation, according to Lewontin, was the finding that crosses between populations from distant localities showed heterosis. This era, 1955 and later, includes the time when I moved from Cold Spring Harbor to Ithaca, New York, and consequently visited Dobzhansky's laboratory much less frequently than I had earlier. Nevertheless, I cannot readily accept Lewontin's assertion. First, interspecific hybrids often exhibit heterosis; this does not mean that species genotypes are not integrated. The coadaptation of species gene pools is revealed by the "medley of unthrifty and . . . sterile types" (Stephens 1946) that the F_1 interspecific hybrids produce. Here it is important to recall that the intermediate fitnesses (measured as egg-to-adult viability) of mixed-population inversion heterozygotes observed by Dobzhansky (Table 2) were not measured among F_1 interpopulation hybrids but among the F_2 recombinant individuals. The theory of coadaptation, in my view, contributes little to either one's expectations or predictions concerning the vigor or health of interpopulation (even interspecific) hybrids. The fate of these hybrids with their two intact genomes must be determined in part by the degree to which transacting gene control signals are discordant and in part by the contradictory timing of events that are under cis-control.

In summary, I feel that Dobzhansky's theory of coadaptation which slowly emerged from his studies of natural populations of *Drosophila pseudoobscura* was sound, although not expressed in general terms. His view of the role played by inversions evolved from one of neutrality, to one admitting of genetic variation between different inversion types (thus allowing local populations to utilize hybrid vigor without then losing it through recombination), to a recognition that the superiority of inversion heterozygotes was itself dependent upon the genetic milieu of the population. By focusing upon the cytologically visible chromosomal inversions, he neglected to consider the attainment of chromosomal monomorphism by a population as possibly being a cryptic coadaptation. In this regard, I think J. C. King and I can take considerable credit for the view (still disputed by many) that the gene pools of local (or isolated) populations tend to become integrated—that is, that events occurring at one gene locus affect those occurring subsequently at other loci.

Acknowledgments

It was both an honor and a pleasure for me to be invited to Leningrad to participate in the international symposium on Theodosius Dobzhansky and the evolutionary synthesis. I wish to thank Professor Inge-Vechtomov, chairman of the organizing committee, for inviting me to participate, Professor Golubovsky for his role in calling me to the attention of the organizing committee, and Professor Mark Adams for many kind and helpful words and deeds.

References

Anderson, W. W. 1969. "Genetics of Natural Populations. XLI. The Selection Coefficients of Heterozygotes for Lethal Chromosomes in *Drosophila* on Different Genetic Backgrounds." *Genetics* 62: 827–836.

Darwin, Charles. 1859. *The Origin of Species*. New York: The Modern Library.

Dobzhansky, Th. 1937. *Genetics and the Origin of Species*. New York: Columbia University Press.

———. 1950. "Genetics of Natural Populations. XIX. Origin of Heterosis

Through Natural Selection in Populations of *Drosophila pseudoob-scura." Genetics* 35: 288–302.

Dobzhansky, Th., and C. Epling. 1944. "Contributions to the Genetics, Taxonomy, and Ecology of *Drosophila pseudoobscura* and Its Relatives." Publication 554. Washington, D.C: Carnegie Institution of Washington.

Dobzhansky, Th., and H. Levene. 1951. "Development of Heterosis Through Natural Selection in Experimental Populations of *Drosophila pseudoob-scura." American Naturalist* 85: 247–264.

Dobzhansky, Th., and O. Pavlovsky. 1953. "Indeterminate Outcome of Certain Experiments on *Drosophila* Populations." *Evolution* 7: 198–210.

Dobzhansky, Th., and M. L. Queal. 1938. "Genetics of Natural Populations. I. Chromosome Variation in Populations of *Drosophila pseudoobscura* Inhabiting Isolated Mountain Ranges." *Genetics* 23: 239–251.

Dobzhansky, Th., and B. Spassky. 1968. "Genetics of Natural Populations. XI. Heterotic and Deleterious Effects of Recessive Lethals in Populations of *Drosophila pseudoobscura." Genetics* 59: 411–425.

Lewontin, R. C. 1974. *The Genetic Basis of Evolutionary Change.* New York: Columbia University Press.

Lewontin, R. C., J. A. Moore, W. B. Provine, and B. Wallace, eds. 1981. *Dobzhansky's Genetics of Natural Populations I–XLIII.* New York: Columbia University Press.

Stephens, S. G. 1946. "The Genetics of 'Corky.' I. The New World Alleles and Their Possible Role as an Interspecific Isolating Mechanism." *Journal of Genetics* 47: 150–161.

Wright, S., and Th. Dobzhansky. 1946. "Genetics of Natural Populations XII. Experimental Reproduction of Some of the Changes Caused by Natural Selection in Certain Populations of *Drosophila pseudoobscura." Genetics* 31: 125–156.

Dobzhansky, Artificial Life, and the "Larger Questions" of Evolution

Charles E. Taylor

"THERE IS grandeur in this view of life ... from so simple a beginning endless forms most beautiful and most wonderful have been, and are being, evolved" (Darwin 1859, p. 490). This grandeur aptly characterizes the worldview of Th. Dobzhansky. Doby believed that evolution permeates our world and that the theory of evolution both provides a unified view of our place in nature that is scientifically well grounded and offers a source of hope for the human future. His life's work was devoted to the development and exposition of that vision.

Although Doby believed that the deepest and most important consequences from the theory of evolution came from its social and philosophical implications, his early work and major contributions were directed more narrowly at the mechanisms of microevolution. He focused on microevolution not because he felt this to be intrinsically the most interesting or most important aspect of evolution but because at the time he entered his scientific career and when he was at his peak of energy and creativity it was these issues that were in greatest need of understanding. In Doby's later years those problems had been largely solved, and he directed his attention toward other issues—especially toward the "emergent" and "creative" power of evolution. He believed that the two most important but poorly understood problems of this sort were the origin of life and the emergence of mind.

Since Doby's death there have been many significant advances in this area. One approach in particular, the study of Artificial Life, seems especially relevant to the fulfillment of his vision. In the pages that follow I will elaborate on Doby's change in emphasis and provide a few examples of research in the field of Artificial Life to illustrate how these relate to the solution of the new problems to which he began to direct his attention.

The Most Important Features of Evolution Are Its Social and Philosophical Implications

The problems that were addressed by biologists in the early decades of this century, and the reasons that it was these problems, not others, are properly the domain of historians of science. I claim no particular expertise in this area and so must rely on the assessment by historians who have studied these issues and on Doby's own words; they are consistent with my experience.

According to Doby in his oral history:

> Evolution is the part of biology which has the highest and most direct implications, the most reflections in fields not immediately connected with biology, such as sociology and philosophy.... This sociological-philosophical angle was really the aspect which interested me most in the whole field of biology, from the earliest days, I think really from my first reading of Darwin at about age fifteen. It is hardly surprising that both during the pre-Revolutionary days in Russia and during the post-Revolutionary days, to biologists, these philosophical-humanistic implications of evolutionism were in the center of attention.
>
> I think it is not an exaggeration to say that probably this interest is what made me, if not a biologist, at least an evolutionist. (Dobzhansky, quoted in Provine 1981, p. 21)

That interest was retained throughout his life.

Although he acquired these personal convictions at a very early age, Doby also recognized that the state of knowledge at that time was insufficient to establish his view scientifically. From this developed his life-long research strategy: he first directed his attention toward understanding natural selection at the micro level, and only afterwards did he take up the larger issues.

> ... from the time of his reading of the *Origin of Species*, what really interested Dobzhansky most about biology were the larger questions of evolution and their meaning for the evolution of humans, human society, and human ethics.... But to understand these larger questions, which he debated endlessly with colleagues and friends in Russia, Dobzhansky believed he needed to understand evolution at the micro-level. Knowledge of evolution of *Coccinellidae* or *Drosophila* was not so much an end in itself for Dobzhansky; it was a means to know more about the most interesting questions, involving the implications for evolution in humans. (Provine 1981, p. 11; see also Provine 1986, p. 329)

It was not until the 1950s that Dobzhansky began writing about social issues and philosophy, and reached full expression in his books *Genetic Diversity and Human Equality* (1973) and *The Biology of Ultimate Concern* (1967).

From Microevolution to "the Larger Issues"

Not only did Doby write little about social, political, and philosophical issues until he was in his fifties, but John Moore, with whom Doby was especially close during his Columbia years, distinctly remembers Doby dismissing such speculation as mere "mental masturbation," a term he sometimes directed at mathematical modeling as well. If Doby had always been interested in the social and philosophical aspects of evolution, why did he not begin to write seriously about them until the 1950s and 1960s?

Part of his change in emphasis may have been a natural accompaniment of aging. It is common for people to become more concerned about philosophy as they pass age fifty to sixty.

Another part of the answer may be that Doby's "glory days" of working with Alfred Sturtevant and Sewall Wright were over by this time, and the excitement of making significant new discoveries was no longer being experienced. Things had soured with Sturtevant many years earlier, so that source of inspiration was gone. Wright withdrew from their collaboration in the late 1940s. The main themes of the "Genetics of Natural Populations" (GNP) papers had already been well developed, and Doby's microevolutionary work by this time was directed largely at the long-term changes of inversion frequencies and in demonstrating the truth of his balance theory of gene variation (see Lewontin 1981).

Although Doby was unquestionably interested in this research, his own views on these were well established. In Ernst Mayr's estimation, the evolutionary synthesis, which had so energized Doby's earlier work, had been largely accomplished by 1947 (Mayr 1980). The unfolding of Doby's vision had required the understanding that accompanied the evolutionary synthesis, but now that this had been achieved it was proper to move onto other things.

I suspect, however, that the most important element in Doby's change in emphasis was Charles Birch. Birch visited Doby's lab about 1950 and then spent a year of field work with him in Brazil during 1955–1956. Later, during a sabbatical leave in 1960, Doby and his wife, Natasha visited Birch in Sydney and spent what Doby later described as "one of the pleasantest, happiest half years of our life." Birch has had an unrelenting dedication to metaphysics and ethics; Doby found him to be one of the most deeply religious people he knew. The two discussed these topics at great length during their association.

Among the circles in which Doby traveled at home in New York, he knew many of the leading figures in philosophy and religion. But Birch was different from these because his training was in biology. He had no formal, professional training in philosophy. Nonetheless, Birch devoted himself to study of the scientific, evolutionary basis of ethics and religion and, among other things, was recently awarded the Templeton Prize, regarded as the highest public recognition for contributions to religion (see Birch and Cobb 1981; Birch 1990). What would have been important were Birch's passion, enthusiasm, and willingness to go forward and speculate about "the larger issues" in spite of lacking formal credentials. This would have given Doby the self-confidence, imprimatur if you will, to "come out of the closet" and speculate on those issues in which he, too, lacked professional qualifications but nonetheless had much interest and much to contribute.

Charles Birch has had a large influence on me as well. In my discussions with Doby about these matters, especially at Mather where Jeff Powell and I worked with him during the summers of his last years, Doby would frequently ask what "our friend Professor Birch" would say about this subject or that. This was done not so much to gain new information, as he probably know Birch's positions better than we did, but rather to provoke a discussion.

This is not to say that Doby and Birch would agree on very much in philosophy. Birch is a committed follower of A. N. Whitehead, who was a speculative philosopher with a strongly evolutionary orientation. Although living in the twentieth century, Whitehead has not too wrongly been characterized as being firmly among the ranks of the great seventeenth-century philosophers (King 1990). Doby, in contrast, was decidedly a romantic. So far as I could discern, his views seemed little different from those debated by the great Russian novelists of the nineteenth century. Although claiming to embrace the writings of Teilhard de Chardin, he was surely not persuaded by any rational argument in Teilhard's writing. Rather he was enchanted by the vision of a progression from inorganic matter up through the great chain of being to humans and ultimately beyond. Doby was certainly familiar with such images even before he came to the U.S. For example, from *War and Peace*:

> In the universe, in the whole universe, there is a kingdom of truth, and we who are now the children of earth are—eternally—children of the whole universe. Don't I feel in my soul that I am part of this vast harmonious whole? Don't I feel that I form one link, one step, between the lower and higher beings, in this vast harmonious multitude of beings in whom the Deity—the Supreme Power if you prefer the term—is manifest? If I see, clearly see, that ladder leading from plant to man, why should I suppose it breaks off at me and does not go farther and farther?" (Tolstoy 1942)

This quotation well characterizes Doby's belief in evolutionary progress, and Teilhard's vision simply provided a moderately respectable authority which Doby could reference. The basic notions were already in Doby's blood long before he took up philosophical speculation. Charles Birch gave him the confidence to articulate and elaborate publicly on them.

The confluence of maturation, association with Charles Birch, and having achieved a satisfactory understanding of the microevolutionary forces were enough for Doby to turn his attention to "the larger issues" that he said had attracted him to the study of evolution in the first place.

THE PROBLEM OF BIOLOGICAL PROGRESS

I have been vague about the meaning of "the larger issues" of evolution up to now. I take these to include a diverse package of concerns: consequences for ethics and social policy; consequences for religion; and consequences for a metaphysical understanding of matter, life, and mind. In Doby's mind these were addressed independently and were only superficially interrelated. Throughout his life Doby painted with a broad brush, paying attention to the most salient issues and leaving the details to others. Sewall Wright once remarked to me that "Dobzhansky was never obsessed with consistency." This was even more true of his concern with "the larger issues" than of his biology, to which Wright was referring.

It seems to me that Doby's success at dealing with the larger issues was quite mixed. For example, his analyses of issues of social policy relating to biological concerns (racism, the IQ controversy, eugenics, and human equality) are models of clarity, sensitivity, and wisdom. These are especially evident in *Genetic Diversity and Human Equality* (Dobzhansky 1973) but can be seen in other writing as well. His religious views, not all of which made it into writing, were very much in the Russian Romantic tradition and as such were quite outside analysis by reason. His metaphysical views, especially in relation to the origin of life and the origin of mind, related to scientific issues and were an important part of the reason that he seemed to embrace the theories of Teilhard. In particular, Doby was attracted to Teilhard's notion of continued evolutionary progress. It was here that he attempted to relate science with philosophy.

Although Doby attacked the problem of progress through evolution with the same zeal, brilliance, and dedication that he employed in his earlier years, this time he was not so successful. Certainly the scientific community has not found Teilhard's arguments very compelling. But that aside, the scientific tools that were necessary for Doby to realize the full scope of his vision of upward progress were not then developed and available to him.

The core problems of progress in relation to biology were unquestionably how and why complexity arises and increases in biological systems (see, e.g., Dobzhansky 1967, p. 19). In particular, Doby thought the origin of life and the origin of humankind's ability to think were the evolutionary problems most in need of explaining; these were best viewed as almost qualitative increases—"evolutionary transcendences," to use his words—in complexity. How to describe this unquestionable direction of evolution except by some guiding force?

The theoretical and mathematical methods for studying complex, nonlinear systems that are necessary for explaining how complexity arises and increases were not available to Doby at the time he was studying those issues. In fact they were not available until computers became more powerful, more widely appreciated, and more utilized. As these tools are now becoming available, a very promising field of research is developing.

This field, called Artificial Life, is attempting to relate the theory of evolution to the origin and growth of biological complexity. Overviews of the field can be found in Langton 1989 and Langton et al. 1991. Here I will briefly describe a few examples drawn from the field which illustrate systems that possess substantial properties of living systems, though at different levels of organization, and attempt to provide some feeling for how and why complexity grows naturally. At the end of the essay I will speculate on the relation of this to the views that Doby held.

ARTIFICIAL LIFE

Artificial life is the study of human-made systems that exhibit behavior characteristic of natural living systems. The hope is that from such study we can gain a better understanding of life itself. Although its roots go back much farther, the field itself is new, dating largely from a meeting at Los Alamos in 1987, and is quite transdisciplinary, with participation from chemists, physicists, molecular biologists, evolutionists, computer scientists, and philosophers. It should be emphasized, however, that the "human-made" aspect of Artificial Life is not critical; it merely reflects the fact that most work in the area so far has been with human-made systems. An alternative characterization is "Complex Adaptive Systems" (Gell-Mann 1990).

There are potentially a very large set of systems which have the "interesting" or "important" properties of life as it has evolved on earth. Those with which we are familiar represent just a small subset of them. Through identifying and characterizing the systems that potentially can or cannot support such patterns, we gain a better understanding of life as it could be and, through this, of life as it is.

The interacting processes that makes up life as we know it on earth involve principally carbon, hydrogen, and oxygen. These elements are present in sugars, nucleic acids, and proteins. But there is no apparent reason why living systems must have this particular composition. The critical feature of life is the process, not simply its material basis. This point has been emphasized by a number of geneticists and evolutionists—e.g., Penrose (in Creswell 1991), Wright (1964), and Birch (1990). "The qualitative novelty of the human estate is the novelty of the pattern, not of its components" (Dobzhansky 1967, p. 58).

What are the "important" or "interesting" properties that characterize life and give natural life its novelty? There is no good, generally accepted definition of "life" that will in all cases distinguish the living from the nonliving. Some have expressed doubt that the distinction is even meaningful, or that a satisfactory definition will ever be found (Minsky 1967). Nonetheless, it is not difficult to list certain properties that most living systems do possess and that most non-living ones do not (Mayr 1982). The most important of these are the ability to reproduce, a high degree of complexity and organization, chemical uniqueness, possession of a genetic program that leads to an ultimate phenotype, history shaped by natural selection, and apparent indeterminacy of actions. But none of these properties by itself is adequate to distinguish living from nonliving.

To my knowledge there are no human-made systems that would be generally accepted as living. A number of systems do, however, possess a substantial subset of the properties we normally attribute to life, listed above. The study and development of those systems is a main focus for research in Artificial Life. Some examples of these include computer viruses (Spafford 1991), autocatalytic systems (Farmer 1991), certain artificially generated ribozymes systems (Robertson and Joyce 1990), and evolving complex processes (Taylor 1991). Here I shall give brief descriptions of some research directed at the two "evolutionary transcendences" that Doby thought best exemplified the achievement of new levels of complexity by nature—the origin of life and the origin of mind (Dobzhansky 1967).

Among the most significant early attempts to characterize the simplest systems capable of supporting life-like behavior (e.g., self-reproduction) was that of Von Neumann (1966), who introduced Cellular Automata in the 1940s for that purpose. (For a lucid and modern account of these, see Wolfram 1986.) Consider a one-dimensional array of compartments, "cells," which can take on integer values (say 0, 1, 2, or 3) at time t. At time t+1 these are all updated, simultaneously, according to some clear rule. An example might be "take the sum of values of the cells immediately to the right and left; if that sum is 3 or more, then assume the value 2; otherwise assume the value 0." It is not difficult to conceive of such systems in two, three, or more dimensions and with more complicated rule systems. Such automata can be shown capable of universal computation and thus able to represent the laws of physical and chemical interactions (Toffoli 1984, Margolis 1984).

To learn which physical and chemical systems are capable of supporting life-like behavior, one might examine all sets of rules in cellular automata and discover which of them can do so. Such an attempt has been made by Chris Langton (1990), building on earlier work by Wolfram (1984). Langton finds that certain rules produce only "frozen" patterns, whereas others produce predominantly chaotic, "gas-like" behavior. It is the rule sets that lie between them in a narrow region, at their phase transition, that are capable of information transmission and pattern reproduction. According to Langton, life is to be only at the edge of chaos.

Once there are systems that are potentially able to support life-like processes, the question arises as to how such processes can arise *de novo*. The critical issue is how self-catalyzing systems of reactions get started and how certain of these can gain local control of their surroundings. Important advances on this question have come from several groups (see Eigen and Schuster 1979, Kauffman 1986, and Farmer et al. 1986). Through such analyses it is possible to identify those stages that the earliest stages of life on earth must have gone through and, presumably, to state the conditions that are necessary for other life-like systems to get established.

Suffice it to say, such analyses have shown it does not appear difficult at all for life-like, self-catalyzing systems to become established. Perhaps this helps explain why life did appear on earth so early in its history.

The UCLA Artificial Life Group, consisting primarily of David Jefferson, Michael Dyer, and me, with our students, has been studying the ecology and evolution of computer processes for several years now, focusing on how they might aid understanding the ecology and evolution of natural organisms and especially how they might give rise to higher levels of complexity in information processing—a prerequisite for the evolutionary transcendence to mind.

Our first system, RAM, was based on the metaphor that living organisms can be viewed as processes, albeit special ones, that organize the matter of which we are made to sustain and propagate that pattern through some period of time. In the RAM world, collections of animal-like processes exist in an environment that also contains a collection of environmental processes; the two interrelate with one another. Associated with each animal process is some memory that may or may not be write-protected (genetic information would be write-protected; length of hair would not). Each process can (1) accumulate knowledge about its environment and about other organisms (i.e., sense, perceive, learn); (2) modify its environment and other organisms (i.e., interact, communicate); (3) calculate (i.e., evaluate, decide, reason); (4) exhibit time-dependent behavior (i.e., age, sense time); (5) change location (i.e., move); (6) create other organisms with mutation or recombination (i.e., reproduce); and (7) terminate or cause others to terminate (i.e., die, kill). RAM is written in Common Lisp and runs on a variety of serial computers (Taylor, Jefferson, Goldman, and Turner 1989; Goldman 1990).

RAM provides a useful way to specify simple rules for individuals that collectively exhibit complicated patterns in groups. It has proven helpful for studying a variety of problems in biology, including symbiosis among marine organisms (Taylor, Muscatine, and Jefferson 1989), and lek formation by sage grouse (Gibson et al. 1990), and for simulating the population dynamics of mosquito control (Fry et al. 1989).

Among the lessons learned from our experience with this program were the importance of a representation for animal programs that naturally permitted evolution and the desirability of a computer architecture that could take advantage of the parallelism inherent in many evolutionary problems (Jefferson et al. 1991).

Our more recent programs—GENESYS/Tracker (Jefferson et al. 1991) and AntFarm (Collins and Jefferson 1991), are designed to run on Connection Machines, made by Thinking Machines Corporation. This computer has many thousands of individual processors, each with their own memories. This makes a natural vehicle for studying the features of emergent properties in natural systems.

The GENESYS/Tracker program consisted of presenting a population of "ant-like" programs with a marked, broken train of 89 squares in a 32×32 toroidal grid. "Ants" were selected and permitted to reproduce on the basis of how many marked squares they had occupied after 200 moves. The population of "ants" typically consisted of 16,000 programs, each occupying one processor on the Connection Machine.

The program that specified an "ant" was either a finite state automaton or an artificial neural net; it was approximately 450 bits long in either event. Depending on its internal state and on the information in the square immediately ahead of it, the "ant" could turn right or left, move one step ahead, or do nothing on each move. That 10 percent of the population that stepped on the most squares during 200 moves were culled and reproduced, with mutation and recombination, to form the next generation.

Reproduction and evolution were achieved by a genetic algorithm (Holland 1975). That is, we began by aligning the bit strings of the two parents and randomly choosing one to begin. For each subsequent bit a random number was examined to determine whether to continue with that string or to "cross over" and copy from the other parent string. After a complete individual was generated, another pass through the genome was made to determine whether or not to "flip" bits, in analogy with mutation. After a new cohort was generated, it was stepped through the same trail as the parents, the top 10 percent selected, and so on for several hundred generations.

A random selection of chromosomes for the starting generation typically led to a mean population score near 0; but after 200 generations the mean score was quite high, typically in the 80s and frequently reaching a perfect score of 89. From random neural networks that scored near 0, the populations evolved to exhibit quite complicated behaviors. A variety of other systems, evolving to accomplish other tasks, including traditional learning, are described in Langton et al. 1991. Whether or not such systems will soon evolve abilities far beyond those of humans, as suggested by Farmer (1991), is not yet certain. In any event, they clearly illustrate the evolution of emergent properties and point toward the creative power of evolution that Doby had found so impressive and difficult to understand with the theoretical, "beanbag" genetics of his day. They also provide some reason to believe that complexity will often continue to increase on its own, giving some notion of progress to evolution.

CONCLUSION

I am quite confident that Doby would have been uncomfortable with many of my assertions above and would have vigorously disputed some of them. The term *Artificial Life* would have been especially provocative for him, but he would not have dismissed it entirely. I recall that soon after I published my first paper (Arnheim and Taylor 1969), one of the first papers on the neutralist theory, Dobzhansky bristled and admonished me, "When I was your age, I, too, believed in such foolishness." In spite of that, he was always encouraging to the young people around him who were seriously exploring new methods or ideas, and he was certainly capable of changing his mind when sufficient evidence had accumulated.

I would like to think, and do in fact believe, that he would be secretly delighted to see how some one-time puzzling issues are becoming resolved and that he would be sharing the optimism that the study of complex adaptive systems is helping to understand how organization has increased, and is increasing, throughout the natural world (Gell-Mann 1990). Doby would have welcomed this as the next step toward fulfilling his vision that evolution provides an all-encompassing view of nature.

Acknowledgments

I wish to thank Daniel Alexandrov and John Moore, whose comments were especially helpful in preparing this paper, and Mark Adams, who went far above and beyond the call of duty while arranging the conference for which the paper was prepared. I am especially indebted to Jeff Powell and to Theodosius Dobzhansky for permitting me to accompany them on their field trips and to join in their philosophizing about Camus, Sartre, . . .

References

Arnheim, N., and C. E. Taylor. 1969. "Non-Darwinian Evolution: Consequences for Neutral Allelic Variation." *Nature* 223: 900–903.

Birch, C. 1990. *On Purpose*. Kensington, UK: New South Wales University Press.

Birch, C., and J. B. Cobb, Jr. 1981. *The Liberation of Life*. Cambridge: Cambridge University Press.

Collins, R., and D. Jefferson. 1991. "AntFarm: Towards Simulated Evolution." In Langton et al. 1991.

Cresswell, H. A. 1991. "Self-reproducing Mechanical Automata." Parts I and II. In C. G. Langton, ed., *Artificial Life II*. Video Proceedings. Redwood City, Calif.: Addison Wesley.

Darwin, C. 1859. *On the Origin of Species*. Cambridge: Harvard University Press.

Dobzhansky, Th. 1967. *The Biology of Ultimate Concern*. New York: New American Library.

———. 1973. *Genetic Diversity and Human Equality*. New York: Basic Books.

Eigen, M., and P. Schuster. 1979. *The Hypercycle: A Principle of Natural Self-Organization*. New York: Springer Verlag.

Farmer, J. D. 1991. "Artificial Life: The Coming Evolution." In Langton et al. 1991.

Farmer, J. D., S. A. Kauffman, and N. H. Packard. 1986. "Autocatalytic Replication of Polymers." *Physica D* 22: 50–67.

Fry, J., C. E. Taylor, and U. Devgan. 1989. "An Expert System for Mosquito Control in Orange County, California." *Bulletin of the Society for Vector Ecology* 2: 237–246.

Gell-Mann, M. 1990. Talk given 9 January 1990 at the Santa Fe Institute, Santa Fe, N. M.

Gibson, R. M., C. E. Taylor, and D. R. Jefferson. 1990. "Lek Formation by Female Choice: A Simulation Study." *Behavioral Ecology* 1: 36–42.

Goldman, S. 1990. "RAM User's Guide." UCLA Cognitive Science Research Program Technical Report. UCLA-CSRP—90–2.

Holland, J. H. 1975. *Adaptation in Natural and Artificial Systems.* Ann Arbor: University of Michigan Press.

Jefferson, D., R. Collins, C. Cooper, M. Dyer, M. Flowers, R. Korf, C. Taylor, and A. Wang. 1991. "The Genesys System: Evolution as a Theme in Artificial Life." In Langton et al. 1991.

Kauffman, S. 1986. "Autocatalytic Replication of Polymers." *Journal of Theoretical Biology* 119: 1–24.

King, A. 1990. "Whitehead, the Modern World, and Artificial Life." Unpublished manuscript.

Langton, C. G., ed. 1989. *Artificial Life.* Reading, Mass.: Addison Wesley.

———. 1990. "Computation at the Edge of Chaos: Phase Transitions and Emergent Computation." Los Alamos National Laboratory Technical Reports. LA-UR–90–379.

Langton, C. G., C. E. Taylor, J. D. Farmer, and S. Rasmussen, eds. 1991. *Artificial Life II.* Reading, Mass.: Addison Wesley.

Lewontin, R. C. 1981. "The Scientific Work of Th. Dobzhansky." In *Dobzhansky's Genetics of Natural Populations I–XLIII,* ed. R. C. Lewontin, J. A. Moore, W. B. Provine, and B. Wallace. New York: Columbia University Press.

Margolis, N. 1984. "Physics-like Models of Computation." *Physica* D 10: 81–95.

Mayr, E. 1980. "Some Thoughts on the History of the Evolutionary Synthesis." In E. Mayr and W. B. Provine, eds., *The Evolutionary Synthesis: Perspectives on the Unification of Biology.* Cambridge: Harvard University Press.

———. 1982. *The Growth of Biological Thought.* Cambridge: Harvard University Press.

Minsky, M. L. 1967. *Computation: Finite and Infinite Machines.* Engelwood Cliffs, N.J.: Prentice-Hall.

Provine, W. B. 1981. "Origins of the 'Genetics of Natural Populations' Series." In *Dobzhansky's Genetics of Natural Populations I–XLIII,* ed. R. C. Lewontin, J. A. Moore, W. B. Provine, and B. Wallace. New York: Columbia University Press.

———. 1986. *Sewall Wright and Evolutionary Biology.* Chicago: University of Chicago Press.

Robertson, D. L., and G. F. Joyce. 1990. "Selection *in vitro* of an RNA Enzyme That Specifically Cleaves Single-stranded DNA." *Nature* 344: 576–578.

Spafford, E. H. 1991. "Computer Viruses—A Form of Artificial Life?" In Langton et al. 1991.

Taylor, C. E. 1991. "Fleshing Out Artificial Life 2." In Langton et al. 1991.

Taylor, C. E., D. R. Jefferson, S. Goldman, and S. R. Turner. 1989. "RAM: Artificial Life for the Exploration of Complex Biological Systems." In Langton 1989, pp. 275–295.

Taylor, C. E., L. Muscatine, and D. R. Jefferson. 1989. "Maintenance and Breakdown of the Hydra-Chlorella Symbiosis: A Computer Model." *Proceedings of the Royal Society, London* B 238: 277–289.

Toffoli, T. 1984. "Cellular Automata as an Alternative to (Rather Than an Approximation of) Differential Equations in Modeling Physics." *Physica* D 10: 117–127.

Tolstoy, L. 1942. *War and Peace.* New York: Simon and Schuster.

Turing, A. M. 1950. "Computing Machinery and Intelligence." *Mind* 59: 433–460.

von Neumann, J. 1966. *Theory of Self-Reproducing Automata.* Completed and edited by A. W. Burks. Urbana, Ill.: University of Illinois Press.

Wolfram, S. 1984. "Universality and Complexity in Cellular Automata." *Physica* D 10: 1–35.

———. 1986. *Theory and Applications of Cellular Automata.* Singapore: World Scientific Publishing Co.

Wright, S. 1964. "Biology and the Philosophy of Science." In W. R. Reese and E. Freeman, eds., *The Hartshorne Festschrift: Process and Divinity.* La Salle, Ill.: Open Court Publishing Co.

PART FOUR

DOBZHANSKY'S WORLDVIEW

The Evolutionary Worldview of
Theodosius Dobzhansky

Costas B. Krimbas

Et in Arcadia ego!

THEODOSIUS DOBZHANSKY (1900–1975) played a key role in the establishment of the modern theory of evolution and the development of population genetics, but he understood his own work in these fields as of broadly humanistic significance. In this essay I will try to present his weltanschauung and its origins, combining bibliographic information together with personal data from a long-standing friendship, which started 1958–1960 when I worked under his guidance as a postdoctoral fellow.

EVOLUTIONARY THEORY

In the 1936 Jesup Lectures at Columbia University and the influential book that resulted from them, *Genetics and the Origin of Species* (1937), Dobzhansky presented the first synthesis of his views on evolution. The appearance of this book marked the birth of the neo-Darwinian, synthetic, or biological theory of evolution; and it was instrumental for the development, in the United States and elsewhere, of experimental population genetics and the experimental study of evolution. It constituted the main stimulus for the appearance of other similar syntheses, those of Mayr (1942), of Simpson (1944), later on of Stebbins (1950), and probably also of Huxley (1942).

Dobzhansky's synthesis, however, was by no means the first exposition of the neo-Darwinian, or synthetic, theory; the works of R. Fisher (1930), J. B. S. Haldane (1932), and especially Sewall Wright (1931) dated one decade earlier, and their papers dealing with that subject earlier still. Actually Dobzhansky explained the mathematical models of Wright in nonmathematical terms and popularized his metaphor of the adaptive field. In his 1937 book and those that followed, as well as in the technical papers forming the "Genetics of Natural Populations" (GNP) series (1938–1975), Dobzhansky borrowed many concepts, technical approaches, and ideas that appear to have come from earlier writings of other evolutionists or geneticists.

Thus, the idea of uncovering hidden genetic variation in natural populations, the raw material on which natural selection operates, is to be found in the 1927 paper of Elena and Nikolai Timoféeff-Ressovsky and ultimately derives from the teachings of their mentor, S. S. Chetverikov. The use of balanced strains for uncovering this concealed genetic variation, rather than inbreeding used by the Timoféeff-Ressovskys, Gordon (1936; Gordon, Spurway, and Street 1939), and Buzzati-Traverso (1942), are to be found in an earlier work of N. P. Dubinin and his collaborators (1934). The idea for phylogenetically ordering triads of gene arrangements differing by overlapping inversions is apparently due to Sturtevant (Sturtevant and Dobzhansky 1936), as is the idea for using lethal genes in order to estimate population size by the frequency of their allelism; in this work as well as others, Wright played an important role in constructing models, suggesting parameters to be estimated experimentally, and estimating them from the experimental data (see Provine 1981, 1986; later on Howard Levene replaced Wright in devising and carrying out statistical treatments). Sturtevant (as Provine has suggested and the transcript of Dobzhansky's oral memoirs confirms) played an important role in guiding the young Dobzhansky in his research, first in cytogenetic and later in populational and evolutionary investigations. According to Provine, the scientific program appearing in the GNP series originated with Sturtevant. Furthermore, Dobzhansky borrowed population cages from L'Héritier and Teissier; the idea of coadaptation derives ultimately from K. Mather; and so on.

Where, then, is the originality of Dobzhansky's approach to be found?

First, even when borrowing ideas, Dobzhansky gave them a new shape, a novel and richer content. Thus, in Dobzhansky's theorizing, coadaptation is a far richer concept than the simple alternate sequence along the chromosome of the two types of alleles encountered in every polygene. Population cages were used to mimic naturally occurring phenomena of selection of inversions; these experiments have led to a deeper understanding of the genetic structure of populations and to concepts like coadaptation, which are not found in L'Héritier's and Teissier's studies concerning the change of frequencies of simple laboratory mutants. Dobzhansky made a judicious use of balanced strains for uncovering the hidden genetic variability; this permitted him and his students (Wallace & Madden 1953) to gather quantitative data regarding all the viability classes of homozygotes for a whole chromosome.

According to Lewontin (in a letter to me dated 27 September 1990) "the most important thing about Dobzhansky was that he took all these different strains and added them to a driving personality so that he made them a living part of evolutionary biology." Lewontin continued: "All the things ... that he was not original in, they were no part of the thinking of biologists and certainly not of anthropologists. What Dobzhansky succeeded in doing was to create the intellectual shape of the field of evolutionary genetics and also have a tremendous impact on beliefs about race and species in anthropology by the shear force of his personality and reemphasis of things that were in the literature as information but not really in people's heads as knowledge."

Second, there are two domains where Dobzhansky was one of the pioneers in establishing dominant trends in the evolutionary thinking of the Anglo-American scientific community.

The first could be described as a deeper understanding of the relations among and between the environment, the genotype, and the phenotype (Dobzhansky 1957; Sinnott, Dunn, and Dobzhansky 1958, pp. 22–25, 29–30). He emphasized the idea that the genotype provides only possibilities and that we are not in the presence of a strict one-to-one relation or of a simple determinism. This is probably a borrowing from Schmalhausen, along with the concept of the norm of reaction.

Related to this way of considering the genetic material is the observation that genes are pleiotropic: each gene is not tied in a one-to-one relation with one phenotypic character but influences the expression of many. Two of Dobzhansky's earlier papers using *Drosophila* dealt with the manifold effects of genes (1924, 1927; at roughly the same time N. Timoféeff-Ressovsky made similar observations). Thus, every character is controlled by many genes. Following this line of thinking to its end: in terms of their phenotypic effects, genes cannot be compared to beans in a bag but should be considered as interacting units. This emphasis on the presence of gene interactions, evident in the concept of coadaptation but also obvious on several other occasions, is directly related to Mayr's concept of the integration and cohesion of the gene pool.

The absence of simple and simplistic mechanical models and the tendency to use more complex ones based on interactions, leading sometimes to unique outcomes, are evident in all Dobzhansky's thinking. He repeatedly characterized selection as a creative process; he was not giving any metaphysical meaning to this creativity—on the contrary, he provided a fair description of the selective processes in nature, noting that they are not simple sieving procedures. Interactions may lead to the formation of novelties, of evolutionary innovations.

> Natural selection works not with genes but with whole genetic endowments; what survives ... is not a gene but a living individual. A gene useful in combination with some genes may be harmful in combination with others. The changes which natural selection promotes at present depend upon the changes that occurred in the past. Natural selection is comparable not to a sieve but to a regulatory mechanism in a cybernetic system. (Dobzhansky 1967c, p. 122)

This ensemble of ideas partially derives from Schmalhausen, as was already pointed out previously, and from Dobzhansky's training as a field naturalist. (Mayr, who expressed similar views, is also a naturalist par excellence.) In the USSR, Dobzhansky's scientific and biological country of origin, this naturalistic tradition was thriving, and thus it is no wonder that the field of phenogenetics flourished there.

The second domain in which Dobzhansky was also a pioneer in the Western scientific community was his species concept. That concept derived from his experience as a field entomologist and ultimately, as Krementsov (1990 and this volume) has pointed out, from the scientific traditions of Russian entomology. Dobzhansky decided that the two races of *Drosophila pseudoobscura* that could not cross in nature were, for that reason, two different species—this despite their near identity morphologically. Therefore, he named "race A" *Drosophila pseudoobscura* and "race B" *D. persimilis*. In this he disagreed with Sturtevant.

Together with Mayr, and according to the ideas already evident in the *persimilis* case, Dobzhansky defended the biological definition of species. In this definition the criterion used to delineate a species is the impossibility of exchanging genes with other similar entities. Species are the most inclusive Mendelian populations; thus the crucial step in cladogenesis amounts to the creation of barriers to gene exchange, in the establishment of isolating mechanisms between natural populations. Dobzhansky was also a pioneer in discovering the genetic basis of isolating mechanisms: his technique of following genetically marked chromosomes in backcross interspecies laboratory hybrids is still widely used in *Drosophila* speciation studies.

In order to explain the tremendous influence Dobzhansky had, another important point should also be stressed: alone or with his collaborators, he produced voluminous experimental results which he speeded to publication. This permitted the formation of a basic literature covering all relevant aspects of a field defined primarily as the experimental illustration of Wright's models, and the experimental study of speciation.

Do these factors exhaust the causes of Dobzhansky's importance and influence? I doubt it. His personality played a key role. He was a very energetic person, full of enthusiasm which he communicated to others, easygoing, and accessible. But he was also exceptional for his time in having a very broad view of the field, of the problems to be tackled, of the way they should be approached. This exceptional broadness of view, together with his personality, permitted him to attract numerous students to his group.

Dobzhansky acquired his views quite early, before emigrating to the United States; with the passing years, he modified and completed them. But his motivation for studying biology, and evolution in particular, were his hope and belief that only in this way could he understand the world and especially humans.

HUMANKIND

Humankind was at all times the prime object of Dobzhansky's interest. It is a pity that he never felt himself a professional, an insider in the field of human genetics or human evolution. In 1964 he published a paper setting forth his broader theoretical views of the subject, entitled "Human Genetics: An Outsider's View" (Dobzhansky 1964a); it was a confession of failure.

Dobzhansky had tried to get into the field during the 1950s. In collaboration with L. C. Dunn, a colleague and close friend who played an important role in Dobzhansky's life, he created an Institute for the Study of Human Variation at Columbia University, New York.

Despite the fact that Dobzhansky was a successful population biologist working with insects, and Dunn was a successful mammalian geneticist, the institute was not a success. An attempt to uncover genetic variation through a study of amino acid contents of the urine using paper chromatography did not produce the expected results; different individuals did show significant differences, but their genetic basis was unclear or, worse, questionable. The institute did not last. In 1965, a decade later, Richard Lewontin succeeded in realizing Dobzhansky's earlier scientific program by the use of electrophoresis. By then, however, it was too late for Dobzhansky to change his field of scientific activities and to work again in human genetics.

However, Dobzhansky did not lose interest in human matters. *Drosophila* experiments were a substitute for human experiments since their results could be projected to people (1967c). In the last decade of his life he performed selection experiments for positive and negative geotropism and phototropism in *D. pseudoobscura* (1967d; Dobzhansky and Spassky 1969; Dobzhansky, Levene, and Spassky 1972; Dobzhansky, Judson, and Pavlovsky 1974). He selected populations monomorphic for different gene arrangements. Then he performed a peculiar experiment, vaguely resembling that of Thoday, in which two populations already selected in contrary directions continued being selected while they underwent a limited exchange of individuals: to a population selected in one direction were added the highest performing individuals in the same direction from the contrary selected population. Dobzhansky then, observed the frequencies of the gene arrangements in the two populations to see whether they retained a difference in their frequencies.

The reasons for performing this experiment become obvious from the familiar names Dobzhansky was giving to these populations: "aristos" and "plebs" (aristocrats and plebeians). From discussions with Dobzhansky, I inferred that he was trying to find out whether a class barrier, permitting a limited exchange of individuals, could sustain a genetic differentiation while different selective pressures were applied to different classes. During his 1960 visit to India, he became fascinated with the complex nature and tightness of the caste and subcaste system. He considered it as a human population genetics experiment on the largest possible scale.

This is not meant to suggest that he favored any class or racial prejudice; on the contrary. And on several occasions he courageously defended the view that people are equal and that racist beliefs are not scientifically justified. This was a recurrent theme in many of his papers dealing with general subjects and many of the lectures he gave to general audiences (1951, 1961a, 1963, 1966a, 1966b, 1967b, 1968a, 1971, 1973a, and 1973b). He believed in the existence of differences between individuals leading to different aptitudes (e.g., musical ability), but he stressed the point that differences do not imply unequal status. He did not believe that the upper classes were biologically superior to the lower classes. He wrote: "Social Darwinism really never had sound biological roots, even though it was, and in some places continues to be, an ideological prop of laissez-faire capitalism" (1962, p. 341). In this he was consistent with his political opinions.

Lewontin has characterized Dobzhansky as a social democrat, a strong proponent of civil liberties, vaguely socialist but certainly left-liberal, and not a great defender of capitalism. He further reports a discussion with him in which Dobzhansky clearly indicated that population genetics had important implications for human life, including politics in the broader sense. In this discussion Dobzhansky apparently declared that he chose this field of investigations because of the implications the research findings had on human affairs: "He said more than once that the study of the human species was the real justification for doing everything else" (Lewontin 1989, p. 21).

As a liberal, Dobzhansky was in general against the application of eugenic measures. He called utopic the positive eugenic proposals of H. J. Muller (a subchapter in Dobzhansky's 1962 *Mankind Evolving* is entitled "Muller's Bravest New World"). Furthermore, as a proponent of a balanced population model, he believed that in many cases genes harmful in one genetic environment or under one specific set of environmental circumstances may be favorable in others. Thus their exclusion from the gene pool was not justified. In some extreme instances, however, when human suffering and the costs to society cannot be overlooked, he proposed to educate and inform the heterozygotes or the affected individuals in the hope that they might freely decide not to have children. In cases of individuals mentally incompetent to reach such a decision, he considered obligatory sterilization (1962, p. 333; 1965).

Interest in humankind, therefore, was his chief motive for studying evolution and population genetics as a consequence, and humankind was the principal object of his concern. Dobzhansky was a humanist; he loved and believed in people; he wanted "nothing humane to remain unknown to him," as the Latin poet expressed it. Because of these interests he traveled widely and visited different countries on all continents.

Until 1942 Dobzhansky refrained from expressing his interest and views on human matters. From this year on, he began writing papers and books that emphasized the subject. *Man* appears in most of the titles of these works. Thus in the 1940s he published four papers plus one book of this sort (Dunn and Dobzhansky 1946); during the next decade, eleven papers and three books (1955, 1956, and Wallace and Dobzhansky 1959); in the 1960s, fifty-eight papers, reviews, and letters and three books (1962, 1964b, 1967a); in the 1970s (during the last six last years of his life, and posthumously) twenty-three papers and two books (1973b, and Dobzhansky and Boesinger 1983). Of a total of fifteen books, then, one is in Russian, one is a genetics textbook, four deal generally with evolution, and nine deal partly or mainly with "man." To these nine books one may perhaps add *The Roving Naturalist* (1980), Glass's compilation of Dobzhansky's letters from his trips around the world.

In most of these books and papers, Dobzhansky presented in a clear way the recent findings of population biology (mainly derived from the experimental work performed with *Drosophila*), explained the mechanisms of evolutionary changes, and indicated their relevance to the study and understanding of humankind. Humans, according to him, held a privileged, a unique position among all organisms. Culture, an adaptive evolutionary product, accelerated the pace of human evolution; evolutionary changes from this time on did not depend on genetic changes, but were transmitted by learning. This in no way meant that natural selection stopped operating on humans: "There exist cybernetic feedback relationships between genetics and culture," i.e., humans were adapting their culture to fit their genes while they were also adapting their genes to their cultural environment (1965). A prerequisite for being able to produce a cultural evolution was the acquisition of a symbolic language, which apparently led to self-awareness. A byproduct of this was also "death awareness," an important characteristic which Dobzhansky repeated with some insistence. Humans were thus aware of themselves, conscious that they evolve, and are eventually able to direct their own evolution.

During the Delos Symposium organized by Constantin Doxiades in 1965, Dobzhansky met Arnold Toynbee, who was presiding over it.

This encounter had some influence on Dobzhansky's thought regarding the typology, life, and evolution of civilizations. He especially liked the idea that civilizations are responses to environmental challenges (1967a, p. 124; Dobzhansky and Boesinger 1968, p. 153). He had already read Toynbee's *Study of History* in preparation for writing his 1962 book, *Mankind Evolving,* but as he confessed to me at that time, he was not very impressed by it (see, however, 1962, p. 17). Toynbee, together with the theologian Paul Tillich, is especially acknowledged in the "Preface ad Hominem" of Dobzhansky's most revealing book, *The Biology of Ultimate Concern* (1967a), where his ideas on evolution, humankind, and religion are crystalized.

Regarding ethics, Dobzhansky favored the approach of Waddington who considered them in an evolutionary perspective. There is a stage in early development in which the child becomes an "authority-acceptor" and an "ethicizing being." The capacity to accept authority and ethical beliefs is an essential adaptation for transmitting from one generation to the next the stored information that constitutes human culture. Dobzhansky was, however, reluctant to accept any criteria for establishing ethical rules based on progress or the promotion of the evolutionary process. On this subject he disagreed with Waddington because, along with Simpson, Dobzhansky found it difficult to decide what progress really meant in this context (1961b; 1962, pp. 343–345).

In their posthumous book *Human Culture* (1983, edited by Bruce Wallace), Dobzhansky and his coauthor Boesinger attempted to formulate an evolutionary explanation of art, the human aesthetic expressions. In 1962 Dobzhansky favored the suggestion of Jenkins to the effect that "man's experience in relation to the world has three functional components—the aesthetic, which focuses upon the 'particularity' (individuality) of things; the affective, which concentrates upon their import; and the cognitive which sees their consequences" (1962, p. 216). Later he put this view aside. Thus he considered art as an evolutionary product and again searched for evolutionary rudiments of aesthetic expressions in animals, in particular lower birds and higher apes, that precede human artistic manifestations.

In the 1983 book Dobzhansky suggested again (as he had in 1962) that all aesthetic values probably stem from sexual selection, as Darwin had proposed. According to this view these aesthetic values may change and evolve: "The idea of an evolution of art and of esthetic sensations is excluded by both ... extreme positions"—that is, on the one hand, the adherence to an established code of beauty, such as that found in art of the classical Greece; or, on the other hand, the rejection of all criteria and all objective value judgments of aesthetic qualities. In place of these alternatives Dobzhansky and Boesinger proposed "an evolutionary conception of art" (1983, p. 114). Their approach to art was also Dobzhansky's approach to religion.

RELIGION AND COSMOLOGY

As his oral memoirs reveal, Dobzhansky was certainly not a materialist, nor was he indifferent to religious matters. There were many priests among his ancestors, and he felt it important to mention this.

In 1963, during our visit to Mount Athos (a visit which, despite several difficulties, he insisted upon making), he received Holy Communion in the Russian Monastery of Haghios Panteleimon. When I asked him what this meant, he replied that it reminded him of his childhood. It was probably an answer meant for me, since he knew I was a nonbeliever.

An interesting episode partly revealing of his thoughts on the matter occurred in Delphi in 1969 during an International Humanistic Symposium. It was

> a kind of Bishop Wilberforce–T. H. Huxley confrontation in miniature.... My role there was to give what I naively believed an uncontroversial account of human evolutionary origins. Not so for the Greek theologians. In the role of Wilberforce was Marcos Siotis, professor of theology, University of Athens. In a letter ... he explains that he felt obliged to rebut the view that mankind evolved from brute animal ancestors, since this view contradicts Genesis.... I ... could only protest in a double capacity, as a scientist and a communicant of the Eastern Orthodox Church. Fortunately, the hidebound rigidity of the Greek section is not shared by the Eastern Church as a whole. Man was and is being created in God's image by means of evolutionary development. (Dobzhansky 1970–1971)

Dobzhansky considered himself to be an active communicant of the Eastern Orthodox Church. According to his last student, Jeff Powell (personal communication), Dobzhansky prayed daily in his final years (by that time he knew he had leukemia and would die).

This needs explication. As a scientist he believed that the Church's tendency to deny widely accepted scientific facts or hypotheses because they seemed to contradict the scriptures showed an unhealthy attitude. Furthermore, he did not consider it wise to search for gaps that science could not at the moment explain, and from these to argue that science was inadequate to explain the world, and to base on this an argument regarding God's existence (1967a, ch. 2). With such arguments, he held, religion would lose battles, as it had lost so many in the past. Instead, he believed that religion should continuously evolve, incorporate scientific findings, and adapt to them.

After 1960 Dobzhansky became interested—and to a certain degree impressed—by the weltanschauung of Teilhard de Chardin, and in 1969 he served as president of the American Teilhard de Chardin Association. Along with Teilhard, Dobzhansky recognized that organic evolution was part of a cosmic process that comprised the birth and evolution of matter and stellar bodies, the appearance and evolution of life, and finally the genesis of humankind. Every time the process passed from one stage of complexity to the next, it transcended itself, first in the transition from matter to life, and then in the genesis of humans, the transition from material life to cultural life. With humankind's appearance, the domain of the noösphere succeeded that of the biosphere. According to Teilhard, evolution had a direction, a privileged axis: from matter, life was born; from life, man; and the process was directing itself to its termination, the Omega point. According to Francisco Ayala (1968), Dobzhansky's student, the Omega point coincides with Saint Paul's notion of the final condition of the world, when all humanity will be united with Christ in a single Mystical Body and everything shall reach consummation in God. I suspect that Ayala played an important role in introducing Dobzhansky to scholastic medieval theology.

In any case, in his writings Dobzhansky modified, or rather adapted, Teilhard's vision according to the dominant neo-Darwinian theory. Thus he dropped Teilhard's claim that evolution proceeds by orthogenesis. Furthermore, he denied that every particular evolutive process had an inherent a priori directionality.

This is evident from his argument concerning extraterrestrial life, an argument he repeatedly presented: If such life existed and evolved independently, it was extremely unlikely that a creature resembling a human would be encountered, contrary to what many believers in the existence of extraterrestrial life expected. Natural selection was short-sighted or, as he preferred to call it, opportunistic; organisms adapted to their environments by natural selection—by increasing the frequency of those genetic variants that *happened to be available* in the gene pool that maximized the fitness of their carriers. This is a blind, short-sighted, and opportunistic process, unique since practically it is not repeatable.

But Teilhard claimed that evolution proceeded by *tatonnement*, that is to say by groping, and Dobzhansky liked this because it gave a good picture of selection, of the absence of a direct goal. He wrote: "Groping is evidently the antithesis of directedness. Particular evolutionary changes, at least on the biological level, show no indication of being, in any meaningful sense, directional. Lack of directionality in particular evolutionary changes does not preclude the possibility that it may be present in general evolution" (Dobzhansky 1968b). Dobzhansky would be inclined to accept that such a tendency, despite exceptions (e.g., in parasitism), exists: we witness the advent of structures of greater complexity, such as humans and their cultural products. "A biologist would go further: a bacterium represents a higher level than a virus, worm higher than bacterium, fish higher than worm, dog higher than fish, and man higher than dog" (1968b).

Of course, Teilhard used a poetic, esoteric language, a prophetic language different from the one used in science. The picture drawn by Teilhard appealed to Dobzhansky, gave a meaning to life—to his life in particular. It helped him escape the anxious feelings that the awareness of death produced. (As M. Ruse kindly informed me, this, together with Russian prayers, is a recurrent theme in the diaries of Dobzhansky's last years.) However, Dobzhansky cautiously refrained from including Teilhard's poetic vision of the Omega point in his own account of what he considered to be the evolutionary process. He avoided metaphysical rhetoric. He never stopped being critical and objective.

Nevertheless, I suspect that since Dobzhansky believed "the evolution of the universe is directional, although not necessarily directed" (1968b), he felt that his duty and the justification was to participate in this evolutionary cosmic process and contribute to its ultimate goal. His contribution was to study evolution, provide a general picture of its mechanisms, and finally to update Teilhard's theses. Thus Dobzhansky saw himself as aiding the evolution of religious thinking in a scientifically evolving world.

Acknowledgments

I would like to thank Drs. K. Gavroglou, R. C. Lewontin, M. Macrakis, J. R. Powell, and B. Wallace who have commented on the manuscript and suggested many ameliorations. In a different form part of it was

presented as an oral communication at the international symposium
on Theodosius Dobzhansky and the evolutionary synthesis in Leningrad in September 1990. I want to thank Drs. M. B. Adams and S. G.
Inge-Vechtomov for inviting me to participate.

References

Ayala, F. J. 1968. "The Evolutionary Thought of Teilhard de Chardin." In A. D.
Breck and W. Yourgrau, eds., *Biology, History, and Natural Philosophy*
(New York: Plenum), pp. 207–216.
Buzzati-Traverso, A. A. 1942. "Genetica di popolazioni di *Drosophila.* I.
Eterozigosi in *Drosophila subobscura* Collins." *Scientia Genetica* 2 (2/3):
190–223.
Dobzhansky, Th. 1924. "Über den Bau des Geschlechtsapparats einiger Mutanten von *Drosophila melanogaster* Meig." *Zeitschrift für induktive Ab-
stammungs- und Vererbungslehre* 34: 245–248.
———. 1927. "Studies on the Manifold Effects of Certain Genes in *Drosophila
melanogaster.*" *Zeitschrift für induktive Abstammungs- und Vererbungslehre* 43: 330–388.
———. 1937. *Genetics and the Origin Of Species.* New York: Columbia University Press.
———. 1951. "Race and Humanity." *Science* 113: 264–266.
———. 1955. *Evolution, Genetics, and Man.* New York: J. Wiley.
———. 1956. *The Biological Basis of Human Freedom.* New York: Columbia
University Press.
———. 1957. "What Is Environment?" *American Naturalist.* 91: 269–271.
———. 1961a. "Human Races." *American Journal of Human Genetics* 13: 349–
350.
———. 1961b. "The Ethical Animal." *Science* 133: 323–324.
———. 1962. *Mankind Evolving: The Evolution of the Human Species.* New
Haven: Yale University Press.
———. 1963. "Genetics of Race Equality." *Eugenics Quarterly* 10: 151–160.
———. 1964a. "Human Genetics: An Outsider's View." *Sympos. O. Biol.* 229:
1–7.
———. 1964b. *Heredity and the Nature of Man.* New York: Harcourt, Brace
and World.
———. 1965. "On Possible Evolutionary Consequences of Different Settle-
ment Patterns." *Ekistics (Athens)* 20: 182–185.
———. 1966a. "A Geneticist's View of Human Equality." *The Pharos of Alpha
Omega Alpha* 29: 12–16.
———. 1966b. "Sind alle Menschen gleich erschaffen?" *Naturwissenschaft
und Medicin* 3: 3–13.
———. 1967a. *The Biology of Ultimate Concern.* New York: New American
Library.

Dobzhansky, Th. 1967b. "On Diversity and Equality." *Columbia University Forum* 10: 5–6.

―――. 1967c. "Of Flies and Men." *American Psychologist* 22: 41–48.

―――. 1967d. "Étude génétique des réactions de Drosophiles à la lumière et à la pesanteur." *Ann. Biol. Clin. (Paris)* 6: 483–497.

―――. 1968a. "On Diversity and Equality." In P. Spackman and L. Ambrose, eds., *The University Forum Anthology.* New York: Columbia University Press.

―――. 1968b. "Teilhard de Chardin and the Orientation of Evolution." *Zygon* 3: 242–258.

―――. 1970–1971. "Perspectives: II. Evolution and Man's Conception of Himself." *The Teilhard Review* 5(2): 65–69.

―――. 1971. "Race Equality." In R. H. Osborne, ed., *The Biological and Social Meaning of Race* (San Francisco: Freeman), pp. 13–24.

―――. 1973a. "Is Genetic Diversity Compatible With Human Equality?" *Social Biology* 20: 280–288.

―――. 1973b. *Genetic Diversity and Human Equality.* New York: Basic Books.

Dobzhansky, Th., and E. Boesinger. 1968. *Essais sur l'Evolution.* Paris: Masson.

―――. 1983. *Human Culture: A Moment in Evolution.* Ed. B. Wallace. New York: Columbia University Press.

Dobzhansky, Th., C. L. Judson, and O. Pavlovsky. 1974. "Behaviour in Different Environments of Populations of *Drosophila pseudoobscura* Selected for Phototaxis and Geotaxis." *Proceedings of the National Academy of Science, USA* 71: 1974–1976.

Dobzhansky, Th., H. Levene, and B. Spassky. 1972. "Effects of Selection and Migration on Geotactic and Phototactic Behavior of *Drosophila.*" *Proceedings of the Royal Society, London* B 180: 21–41.

Dobzhansky, Th., and B. Spassky. 1969. "Artificial and Natural Selection for Two Behavioral Traits in *Drosophila pseudoobscura.*" *Proceedings of the National Academy of Science, USA* 62: 75–80.

Dubinin, N. P., and fourteen collaborators. 1934. "Experimental study of the ecogenotypes of *Drosophila melanogaster*" (in Russian). *Biologicheskii Zhurnal* 3: 166–216.

Dunn, L. C., and Th. Dobzhansky. 1946. *Heredity, Race, and Society.* New York: New American Library.

Dunn, L. C., and Th. Dobzhansky. 1952. *Heredity, Race, and Society,* 2d ed. New York: New American Library.

Fisher, R. A. 1930. *The Genetical Theory of Natural Selection.* Oxford: Oxford University Press.

Glass, B., ed. 1980. *The Roving Naturalist: Travel Letters of Theodosius Dobzhansky.* Philadelphia: American Philosophical Society.

Gordon, C. 1936. "The Frequency of Heterozygosis in Free-living Populations of *Drosophila subobscura.*" *Journal of Genetics* 33: 25–60.

Gordon, C., H. Spurway, and P. A. R. Street. 1939. "An Analysis of Three Wild Populations of *Drosophila subobscura*." *Journal of Genetics* 38: 37–90.

Haldane, J. B. S. 1932. *The Causes of Evolution*. London: Longmans.

Huxley, J. S. 1942. *Evolution: The Modern Synthesis*. London: Allen and Unwin.

Krementsov, N. L. 1990. Oral communication made at the international symposium on Theodosius Dobzhansky and the evolutionary synthesis, Leningrad, September 1990.

Lewontin, R. C. 1989. Notes on Th. Dobzhansky. Unpublished manuscript.

Lewontin, R. C., J. A. Moore, W. B. Provine, and B. Wallace, eds. 1981. *Dobzhansky's Genetics of Natural Populations I–XLIII*. New York: Columbia University Press.

Mayr, E. 1942. *Systematics and the Origin of Species*. New York: Columbia University Press.

Provine, W. B. 1981. "Origins of the 'Genetics of Natural Population' Series." In R. C. Lewontin, J. A. Moore, W. B. Provine, and B. Wallace, eds., *Dobzhansky's Genetics of Natural Populations I–XLIII* (New York: Columbia University Press), pp. 1–76.

Provine, W. B. 1986. *Sewall Wright and Evolutionary Biology*. Chicago: University of Chicago Press.

Simpson, G. G. 1944. *Tempo and Mode In Evolution*. New York: Columbia University Press.

Sinnott, E. W., L. C. Dunn, and Th. Dobzhansky. 1958. *Principles of Genetics*, 5th ed. New York: McGraw-Hill.

Stebbins, G. L. 1950. *Variation and Evolution In Plants*. New York: Columbia University Press.

Sturtevant, A. H., and Th. Dobzhansky. 1936. "Inversions in the Third Chromosome of Wild Races of *Drosophila pseudoobscura*, and Their Use in the Study of the History of the Species." *Proceedings of the National Academy of Science, USA* 22: 448–450.

Timoféeff-Ressovsky, H., and N. W. Timoféeff-Ressovsky. 1927. "Genetische Analyse einer freilebenden *Drosophila melanogaster* Population." *Roux Archiv für Entwicklungsmechanik der Organismen* 109: 70–109.

Wallace, B., and Th. Dobzhansky. 1959. *Radiation, Genes, and Man*. New York: Holt.

Wallace, B., and C. Madden. 1953. "The Frequencies of Sub- and Supervitals in Experimental Populations of *Drosophila melanogaster*." *Genetics* 38: 456–470.

Wright, S. 1931. "Evolution in Mendelian Populations." *Genetics* 16: 97–159.

Dobzhansky and the Biology of Democracy:
The Moral and Political Significance
of Genetic Variation

John Beatty

"WHAT CAN science do for democracy?" It was a question raised often in the United States in the late 1930s and 1940s. Prompting it was a serious threat to democracy—the rise and spread of fascism, combined with a nagging suspicion (not without grounds) that American science had in the past served more to undermine the cause of democracy than to advance it. In other words, what—for a change—could scientists do to support democracy in that trying time?

Among the scientists who posed the question and tried to answer it was the barely American, Russian émigré Theodosius Dobzhansky. In this paper, Dobzhansky serves to illustrate the political activity—in the form of science—of many American scientists in response to the rise of fascism and in reaction to the threateningly antidemocratic agendas and messages of earlier science.

Dobzhansky is an especially interesting case because his main area of research—genetic variation and its role in evolution—had traditionally inspired so much skepticism about, and downright hostility toward, democratic ideals. By looking closely at Dobzhansky's work, it is also possible to see the difficulties facing a common and otherwise understandable view of the political agendas of American scientists at the time. It is often argued, of course, that American scientists responded to the rise of fascism by defending the intellectual freedom and political autonomy of science. But in fact scientists like Dobzhansky were heavily into broader political ideology—the ideology of democracy. As Gregg Mitman suggests in his recent, elegant studies of American biology, in the interwar years the American biologist came forward as "democracy's greatest savior."[1]

---

[1] Mitman 1990, p. 463; see also Mitman 1988 and forthcoming.

This may seem an odd perspective on Dobzhansky; it is, after all, a very American context in which to place a very Russian thinker. Most Anglo-American historians who discuss Dobzhansky treat him in effect as an American, but mostly by virtue of unintentionally ignoring the details of his Russian heritage. (I have certainly been guilty of this; the obvious exception, of course, is Mark Adams.) Here, however, I am quite consciously (and thus perhaps even more foolheartedly!) trying to portray the distinctly American context of Dobzhansky's work.

Because of my emphasis on Dobzhansky's more immediate American context, my account of his views is certain to be incomplete; it is hard to believe that Dobzhansky's scientific/political attitudes and pursuits in the late 1930s and thereafter do not also reflect in important respects his Russian background. I do have a few—but only a few—suggestions in this regard. I hope more information will emerge concerning Dobzhansky's early years that will throw much more light on this matter.

ARISTOCRACY VERSUS DEMOCRACY

In order to make sense of the supposed challenges to democracy posed by science in the late nineteenth and early twentieth centuries, it is important to point out the prevalence of naturalistic thought at the time (Persons 1958). The so-called naturalistic outlook was opposed to supernaturalistic, religious, metaphysical, or otherwise aprioristic rationalizations. More positively (if somewhat circularly) construed, the naturalistic outlook was "scientific." For example, naturalists scorned religious justifications of and dogmatic adherences to political systems, calling instead for scientific assessments.

Democracy in particular came under considerable scrutiny by naturalistically inclined critics (see especially Persons 1958 and Purcell 1973). In particular, hereditary differences were invoked in various ways to criticize democratic notions of equality. Both the tone and substance of this challenge are captured in the following passage from the sociologist and political economist William Graham Sumner:

> Mr. Sinclair says that "Democracy is an attitude of soul. It has its basis in the spiritual nature of man, from which it follows that all men are

equal, or that, if they are not, they must become so." Then Democracy is a metaphysical religion or mythology. The age is not friendly to metaphysics or mythology, but it falls under the dominion of these old tyrants in its political philosophy. If anybody wants to put his soul in an attitude, he ought to do it. The "system" allows that liberty, and it is far safer than shooting. It is also permitted to believe that, if men are not equal, they will become so. If we wait a while they will all die, and then they will all be equal, although they certainly will not be so before that. (*Essays*, vol. 2, pp. 126–127)

The historical literature on findings of genetic "inferiority" and "superiority," and their supposed political relevance, is extensive (see, e.g., Ludmerer 1972 and Kevles 1985). Instead of trying to summarize it, I will just point out a few ways in which hereditary differences were invoked to criticize democracy.

It was commonly argued that genetic diversity favors aristocracies over democracies, or at least more aristocratized forms of democracy. For instance, Sumner argued that differences "by birth" blur the very distinction between aristocracy and democracy:

Aristocratic and democratic are indeed currently used as distinctly antagonistic to each other, but whether they are so or not depends upon the sense in which each of them is taken, for they are words of very shifting and uncertain definition. It is aristocratic to measure men and scale off their social relations by birth; it is democratic to deny the validity of such distinctions and to weigh men by their merits and achievements without regard to other standards. In this sense, however, democracy will not have anything to do with equality, for if you measure men by what they are and do, you will find them anything but equal. This form of democracy, therefore, is equivalent to aristocracy in the next sense. For, second, aristocracy means inequality and the social and political superiority of some to others, while democracy means social and political equality in value and power. But no man ever yet asserted that "all men are equal," meaning what he said. (Sumner 1896–1897, in *Essays*, vol. 2, p. 317)

Sumner and many others argued that democratic representation, in the extreme, was inappropriate given extensive genetic differences in mental and moral sensibilities because intrinsically less capable voters will choose intrinsically less capable representatives and will insist on having a voice in matters that they are intrinsically incompetent to judge (e.g., *Essays*, vol. 2, pp. 304–359 and 260–265). The geneticist Edward East argued in this regard that "our whole governmental system is out of harmony with genetic common sense" (1929, p. 300). Geneticists Paul Popenoe and Roswell Johnson claimed that scientific expertise (such as their own expertise on eugenic issues) would never be properly appreciated in a democracy—in other words by the less competent majority (1923, pp. 360–362).

Criticisms such as these were not intended to undermine democracy completely but rather to circumscribe it in such a way as to give more privileges and responsibilities to the (genetically) most capable citizens. Thus East argued that it would not be completely "undemocratic" to weight the votes of those who pass special competency requirements:

> Undemocratic? Nonsense! The scheme need not be undemocratic. Let every child, without regard to its race, colour, or economic station, receive all the education it is capable of assimilating at the expense of the State. Provide educational facilities of all kinds—Latin schools, English schools, business schools, trade schools, technical schools, colleges and universities; and give every citizen of the republic the opportunity to develop all the ability he possesses. Then extend or curtail the franchise in accordance with the outcome of the probationary test. Plural voting privileges for those who make good, provided all have equal opportunities, might be the height of wisdom; but apart from this heretical suggestion, the plan has the merit of being just, reasonable, and truly democratic. (East 1929, p. 301)

Similarly, Popenoe and Johnson argued that it was not democratization per se that was the problem but rather "too great democratization." The best democracy is partly aristocratic:

> The wise democracy is that which recognizes that officials may be effectively chosen by vote, only for legislative offices; and which recognizes that for executive offices the choice must be definitely selective, that is, a choice of those who by merit are best fitted to fill the positions.... Good government is then an aristo-democracy. (Popenoe and Johnson 1923, pp. 361–362)

There were geneticists who denied that findings in their field supported the aristocratization of American democracy, but the point is that such reasoning was common enough to merit rebuttal. For instance, Herbert Spencer Jennings attacked "the fallacy that biology requires an aristocratic constitution of society" (1930, pp. 220–221). Such a fallacy would not have been worth dispelling if it had not been commonly committed.

The Good of the Individual versus the Good of the Group

Another challenge to democracy posed by genetic diversity had to do with the way in which hereditary differences were supposedly (1) exacerbated and at the same time (2) rationalized, in terms of their role in the process of evolution by natural selection. For instance, in connection with the first point, Sumner argued that democratic egalitarianism had become less and less feasible in the United States as the West had been conquered, as extra resources had diminished, as the struggle for existence, including competition for scarce resources, had intensified, making even greater losers of the genetically inferior (e.g., *Essays*, vol. 2, pp. 92–95, 293–294, 310–311).

In connection with the second point, the struggle for existence, with its attendant losers (as well as winners), was considered to be good, not of course for the losers but for the race or for humankind. The progressive evolution of races and species depend on it. This constituted a "Darwinian" rationalization of the genetic differences that supposedly underlay so much inequality and of the consequent suffering of the losers in the struggle for existence.

The source of this type of reasoning was the source of Darwinism itself. Just as Malthus justified the individual losses due to competition for scarce resources in terms of the general social advancement that (supposedly) ensued, so, too, Darwin viewed the individual casualties of the struggle for existence in the rest of the animal and plant world as means to a generally greater end. Trying to render his mechanism of evolution as palatable as possible, Darwin advised that, "When we reflect on this struggle, we may console ourselves with the full belief, that the war of nature is not incessant, that no fear is felt, that death is generally prompt, and that the vigorous, the healthy, and the happy survive and multiply" (1859, p. 79). Individual human losses were more painful and regrettable, but still justifiable, as Darwin argued in *The Descent of Man*:

> Man, like every other animal, has no doubt advanced to his present high condition through a struggle for existence consequent on his rapid multiplication; and if he is to advance still higher he must remain subject to a severe struggle. Otherwise he would soon sink into indolence, and the more highly-gifted men would not be more successful in the battle of life than the less gifted. Hence our natural rate of increase,

though leading to many and obvious evils, must not be greatly diminished by any means. There should be open competition for all men; and the most able should not be prevented by laws or customs from succeeding best and rearing the largest number of offspring. (Darwin 1871, p. 403)

This sort of biological argument for considering the welfare of the individual subordinate to the welfare of the group took various political manifestations during the late nineteenth and early twentieth centuries; these ranged from the more laissez-faire version often dubbed "social Darwinism" to the more interventionist form known as "eugenics."[2] But whether by natural selection or artificial selection, improvement of the group was seen to depend inevitably upon individual differences in worth. Thus, arguing for the virtues of a laissez-faire economy, Sumner reasoned that,

We can take the rewards from those who have done better and give them to those who have done worse. We shall thus lessen the inequalities. We shall favor the survival of the unfittest, and we shall accomplish this by destroying liberty. Let it be understood that we cannot go outside of this alternative: liberty, inequality, survival of the fittest; not-liberty, equality, survival of the unfittest. The former carries society forward and favors all its best members; the latter carries society downwards and favors all its worst members. (*Essays*, vol. 2, p. 94, see also pp.125, 152–153)

And as the eugenicists Popenoe and Johnson reasoned,

. . . this philanthropic spirit, this zealous regard for the interests of the unfortunate, which is rightly considered one of the highest manifestations of Christian civilization, has in many cases benefited the few at the expense of the many. The present generation, in making its own life comfortable, is leaving a staggering bill to be paid by posterity.

It is at this point that eugenics comes in and demands that a distinction be made between the interests of the individual and the interests of the race. It does not yield to any one in its solicitude for the individual unfortunate; but it says, "His happiness in life does not need to include leaving a family of children, inheritors of his defects, who if they were able to think might curse him for begetting them and curse society for allowing them to be born." (Popenoe and Johnson 1923, pp. 150–151)

[2] Recently historians have emphasized the differences over the similarities between social Darwinism and eugenics, stressing the laissez-faire aspects of the former and the government interventionist aspects of the latter. For the purposes of this paper, though, the similarities of social Darwinism and eugenics outweigh the differences.

MORAL MISGIVINGS

This good-of-the-group perspective on individual differences in worth is the best entrée into Dobzhansky's involvement with issues concerning the relevance of genetic variation for democracy. Dobzhansky was well aware of the problems raised by that line of reasoning, if not because he discovered them himself then because his attention was drawn to them by friends and colleagues.

The qualms of the geneticist T. H. Morgan are well known. In response to Darwin's advice that we keep in mind how often death is prompt, etc., Morgan replied,

> The kindliness of heart that prompted the . . . sentence may arouse our admiration for the humanity of the writer, but need not, therefore, dull our criticism of his theory. For whether no fear is felt, and whether death is prompt or slow, has no bearing on the question at issue—except as it prepares the gentle reader to accept the dreadful calamity of nature, pictured in this battle for existence, and make more contented with their lot "the vigorous, the healthy, and the happy." (Morgan 1903, p. 116)

Over time Morgan came to see more empirical if not any more moral virtue to Darwin's theory. He could then defend evolution by natural selection, but not without warning the reader that, "We must be on our guard . . . lest continuous attention to the marvels of nature direct our thoughts away from the other side of the picture—the wastefulness, the cruelty, the tragedies of nature, as they appear to us" (Morgan 1932, p. 117).

Morgan discussed such misgivings with Dobzhansky, who had traveled to the United States in 1927 on a Rockefeller Foundation fellowship to work with Morgan. He stayed on in Morgan's lab at the California Institute of Technology until 1939, having decided in the interim to apply for U.S. citizenship; he left to accept a professorship at Columbia University. Dobzhansky was the most evolutionarily minded member of Morgan's research group and an important resource to Morgan in this regard. The two men also had many discussions about broadly "philosophical" issues surrounding evolution.[3]

---

[3] Morgan singled out Dobzhansky for special thanks in the preface to his 1932 book, *The Scientific Basis of Evolution*, in which he also discussed his moral reservations about Darwinism. See also Dobzhansky's oral memoirs (1962–1963) pp. 350–352, and Allen, in this volume.

Dobzhansky had already encountered similar concerns in Russia. As Daniel Todes has shown (1987, 1989), there was in the late nineteenth and early twentieth centuries in Russia a tradition of evolutionary theorizing that was self-avowedly "Darwinian" but with the Malthusian elements deemphasized or altogether excised. Those Russians Todes refers to as "non-Malthusians" found Malthus's views of competition empirically unsatisfactory; the "anti-Malthusians" found those views to be, in addition, morally repugnant. From both points of view, the Malthusian elements of Darwin's theory were considered to reflect better the moral and political ideals of Darwin's class and culture than the state of nature.

Dobzhansky often remarked with regard to his Russian heritage that, where he had come from, evolution was a "philosophical" and not just a scientific topic (Dobzhansky 1962–1963). And as we see, it did not cease to be a philosophical topic for him during his early years in the United States with Morgan, during which time he conceived and initiated his long-term research project on the evolutionary genetics of natural populations, and during which time he also wrote *Genetics and the Origin of Species* (1937), which was perhaps the most influential document of the evolutionary synthesis.

Central to Dobzhansky's research and to his conception of evolution, from this period on, was a moral concern about the extent to which the evolutionary well-being of the population or species depends upon the differential welfares of its members. This concern intersected and was reinforced by concurrent and more general concerns about the relationship between genetic diversity and democratic egalitarianism.

Dobzhansky referred to the conflict between the welfare of a population and the welfare of its members as a "paradox." It was a paradox he sought to resolve in numerous ways, by emphasizing differences between individuals, all the while downplaying any resultant inequalities.

THE "PARADOX"

Ultimately the problem had to do with the Janus-faced character of mutation. Random mutation was widely considered to be, almost of necessity, deleterious. That is, assuming that species are already well adapted by natural selection to their environments, random mutations would seem far more likely to detract from, rather than to improve, the fitness of their possessors. It was common to invoke the analogy of random tinkering with a fine-tuned watch: wouldn't that be much more likely to hamper than to enhance its time-keeping function (e.g., Muller 1929, p. 488)? This theoretical, almost a priori argument for the general deleteriousness of mutation was supported by empirical studies of mutation—e.g., by the Morgan group, and especially by H. J. Muller's work on radiation-induced mutations.[4]

Not only are mutations generally deleterious, they are also (appropriately enough) infrequent—mutation rates were calculated on the order of 10^{-3}–10^{-6}. But the infrequency, combined with the general harmfulness of mutation, presented a problem. Presumably environments change from time to time, leaving previously well adapted species less well so. To adapt quickly, species need a substantial amount of variation for selection to act upon. But the harmfulness of mutations in the previous environment has been an obstacle to storing them, and they are not readily available *de novo*.

Population geneticists such as Sergei Chetverikov (1961 and 1929?), R. A. Fisher (1930), and Sewall Wright (1931) acknowledged this problem and set out to solve it—i.e., they endeavored to show how species store the variation needed for further evolution by natural selection, even though mutations arise infrequently and are generally deleterious for their possessors.[5] They offered quite different explanations. But the similarities in the problems they posed outweighed the differences in their solutions, especially when compared with Muller, for whom mutation was not so Janus-faced. For Muller, the accumulation of mutations within a species could only be bad. Fortunately, he believed, selection is constantly eliminating mutations. There was, in his opinion, only one species—namely, our own—in which mutations accumulate to any very great extent. Through the amenities of civilization, he argued, we have managed to escape natural selection, to our ultimate detriment. Only through a conscious eugenics program can we now successfully reverse the rising tide of

[4] Muller argued that radiation-induced point mutations were not intrinsically different—certainly no more deleterious than spontaneous mutations. But radiation certainly induced a lot of viability-reducing mutations. Of course, that is precisely how Muller's ClB technique detected mutation—i.e., by its total negative impact on viability (see, e.g., Muller 1934).

[5] This list of concerned population geneticists is meant to be exemplary, not complete. I am grateful to Vassily Babkov for providing me with a copy of Chetverikov's unpublished excerpt, "Mutations in Nature." It is in this section that Chetverikov addresses the problem described in the text.

deleterious mutations that will otherwise result in the extinction of our kind (see, e.g., Muller 1950a and 1950b).

Dobzhansky was one of those who, unlike Muller, and more like Chetverikov, Fisher, and Wright, saw a serious problem in the conflicting disadvantageous–advantageous aspects of mutation. The problem played a central role in the first edition of his *Genetics and the Origin of Species* (1937).

There Dobzhansky referred to the problem as a "paradox of viability": that is, it is a paradox that the viability of a species depends on the possession of genetic variation that decreases the viability of many species members. Of course, that would really only be a contradiction if one assumed that nature is not so preferential. Indeed, to the extent that the problem is real, Dobzhansky claimed, it represents an "imperfection of nature." As he summarized the problem, "A species perfectly adapted to its environment may be destroyed by a change in the latter if no hereditary variability is available in the hour of need. Evolutionary plasticity can be purchased only at the ruthlessly dear price of continuously sacrificing some individuals to death from unfavorable mutations. Bemoaning this imperfection of nature has, however, no place in a scientific treatment of the subject" (Dobzhansky 1937, pp. 126–127).

Of course, "bemoaning" the state of nature in a "scientific treatment of the subject" was precisely what Dobzhansky was doing. Elsewhere, around the same time, he remarked in similarly self-contradictory fashion, "The general picture of evolution thus arrived at will certainly be far from pleasing to those who regard nature as an embodiment of kindness. The writer must confess that this picture is not pleasing to him either. The words 'good' and 'bad' are not to be found, however, in the scientific lexicon" (Dobzhansky 1938, p. 449).

The moral concerns did not fade with time. In his influential book *Mankind Evolving*, published in 1962, Dobzhansky stated the same problem, though more succinctly: "Adaptation involves genetic elimination of countless ill-adapted variants. The welfare of the species is paid for by the misery of many individuals" (p. 335).

Dobzhansky pursued this problem mainly in connection with investigations of genetic variation in natural populations of *Drosophila*, very much in the tradition of Chetverikov, whose work Dobzhansky followed from afar. But it was the bearing of the problem on the condition of humankind that concerned him most. And just as Muller did not hesitate to draw sociopolitical lessons from his

Drosophila researches, Dobzhansky was quick to do the same. Indeed, he had aimed his discussion of the problem in *Genetics and the Origin of Species* directly at "those eugenical Jeremiahs" who overlook the importance of evolutionary plasticity when they (mistakenly) extrapolate from the general deleteriousness of mutations for individual humans to the general deleteriousness of mutations for humankind (Dobzhansky 1937, p. 126).

Dealing with this paradox of viability—this imperfection of nature—especially as it pertained to humans, thus became one of Dobzhansky's central research agendas. In part he was concerned to show that genetic variation was indeed very important to humankind—that the "eugenical Jeremiahs" were wrong. In part he was also concerned to resolve the paradox—to show that variation was not always or necessarily bad for individuals.

OCCASIONS FOR RESPONSE

The central role and moral tone of Dobzhansky's discussion of the problem in *Genetics and the Origin of Species* and elsewhere support later reminiscences to the effect that *Drosophila* work had always been, for him, a means to human ends. As he further recollected, the period surrounding the publication of the first edition of *Genetics and the Origin of Species* was especially significant in this regard. "It was consequently about 1937 or '38—that was in my life as a researcher crucial—I became convinced that here is a chance of discovering something both of general biological interest, and something which would have bearing and relevance to man and to the human condition. We were playing with flies, but the existence of this concealed genetic variability, or of the genetic load, is ... [a more general problem affecting humans as well]" (Dobzhansky 1962–1963).

The problem became acute for Dobzhansky at this time in part because it was also at this time that his research into the genetics of natural populations of *Drosophila* first began to yield results demonstrating quite large amounts of genetic variation. Dobzhansky also began around this time to pursue various generalizable mechanisms to account for the maintenance of this variation, in other words to show why in general one could expect to find sufficient genetic variation to ensure the evolutionary plasticity of populations and species. I will not consider in this paper all of the mechanisms he entertained and all the evidence he marshaled in support of his various positions (see Lewontin 1981 and Beatty 1987). My concern here is more with Dobzhansky's attempts to deal with the moral and political implications of extensive genetic diversity.

The individual/group conflict also grew in importance for Dobzhansky in the late 1930s because of a variety of other related circumstances including the rise of fascism, the rise of political activism among scientists aimed at combating fascism and preserving democracy, and Dobzhansky's increasing contacts with and ultimately his move to Columbia University, which was a center of scientific political activism—and in the midst of all that, and presumably making him all the more politically self-conscious, Dobzhansky receiving his U.S. citizenship!

Dobzhansky spent most of the fall semester 1936 at Columbia presenting the prestigious Jesup Lectures, which were the basis for *Genetics and the Origin of Species*; as I mentioned earlier, he moved to Columbia in 1939. As Peter Kuznick (1987) has made clear, Columbia was in the late 1930s and early 1940s a "hotbed of antifascist sentiment" with scientists leading the charge. Most prominent among the scientific activists at Columbia was the anthropologist Franz Boas. But many other accomplished and influential Columbia scientists were also involved, including the geneticist Leslie Dunn, who became one of Dobzhansky's closest friends.

Boas and Dunn were among the leaders of the Columbia University Federation for Democracy and Intellectual Freedom; they also served on the executive committee of the nationwide American Committee for Democracy and Intellectual Freedom. These groups represented American science in the fight against fascism and the defense of democracy. As Boas argued the cause, "The present outrages in Germany have made it all the more necessary for American scientists to take a firm antifascist stand.... [W]e scientists have the moral obligation to educate the American people against all false and unscientific doctrines, such as the racial nonsense of the Nazis. The agents of fascism in this country are becoming more and more active, and we must join with all men of goodwill in defending democracy today if we are to avoid the fate of our colleagues in Germany, Austria, and Italy" (quoted in Kuznick 1987, p. 187). Similarly, Dunn addressed such topics as "Natural Science and Democracy," in this case in a radio address on Armistice Day 1937, where he raised the question "What can science do for democracy?" (Dunn, in Ludmerer 1972, p. 130).

Among the areas of science that seemed relevant to the respective fates of fascism and democracy was genetics. Ideas of genetic superiority and inferiority seemed to be part and parcel of fascism. Correlatively, it was often argued, insofar as Americans had defended the idea of genetically superior and inferior types (in other words, to that considerable extent), we had also reasoned like fascists (e.g., Dunn in Ludmerer 1972, p. 130). What could American geneticists now say about genetic differences that might be more in keeping with democratic ideals of equality?

Dobzhansky's attempts to resolve the evolutionary conflict between the welfare of a population or species and that of its members should be understood in part in the context of this period in which so many American scientists—especially this self-consciously new American scientist—tried so hard to support democratic ideals. It became even more pressing for Dobzhansky as he found ever more evidence of ever more genetic diversity, and as he sought to explain why such high levels of diversity were to be expected.

Although it is somewhat of an aside, it is important to point out that at this time democracy was being promoted vigorously not just in response to fascism but more generally in response to totalitarianism, including Soviet communism. Dobzhansky did have colleagues at Columbia who used biology to defend democracy against totalitarianism in general. For example, Edmund Sinnott, who coauthored a popular multiedition genetics textbook with Dunn (1925, 1932, and 1939), and later with both Dunn and Dobzhansky (1950 and 1958), argued that genetic diversity promotes democracy by undermining "totalitarian attempts to regiment us into an army of standardized robots who would march off the assembly lines of indoctrination as monotonously alike as a string of jeeps" (Sinnott 1945–1946, p. 66).

But neither Dunn nor Dobzhansky themselves defended democracy against communism. This is not so surprising in the case of Dunn, who for instance also served in important capacities on the National Council of American-Soviet Friendship. It may seem more surprising in the case of Dobzhansky since we know he was not supportive of the October Revolution and was pro-White during the Russian Civil War (Dobzhansky 1962–1963).

Dunn and Dobzhansky defended democracy mainly against fascism and aristocracy. Why aristocracy? Recall that American biologists had previously defended aristocracy versus democracy. Aristocracy was thus every bit as much a foil for prodemocracy American biologists in the 1930s and 1940s as were fascism and communism. It is surely also relevant here that while Dobzhansky was opposed to the October Revolution, he was very much in support of the February Revolution ending the rule of the tsars. I realize this is merely suggestive, but I believe that aristocracy continued to represent a much more pernicious political system from Dobzhansky's point of view than did communism.

Attempted Resolutions

It is worth contrasting the sort of resolution between democracy and genetic variation that Dobzhansky sought with the sort of "resolution" offered, for example, by the geneticist D. C. Rife in his 1945 book, *The Dice of Destiny: An Introduction to Human Heredity and Racial Variation*, which Dobzhansky reviewed (see esp. pp. 147–156). Like Dobzhansky, Rife insisted on the pervasiveness of genetic variation, and like Dobzhansky, Rife insisted that genetic variation was consistent with democracy. Nonetheless, like so many earlier thinkers, Rife believed that the greater the genetic variation, the harder it was to maintain a democracy. For Rife, democracy had to be promoted in spite of genetic variation. In the years to come Dobzhansky would seek more positive connections between genetic diversity and democracy.

In 1946 Dobzhansky and Dunn teamed up to write the short, popular *Heredity, Race, and Society*. There they posed the rhetorical demand for their services: "Now ... you hear people say 'There ought to be a scientific solution,' or 'science ought to be able to tell us how to live in a world where ... human differences keep stirring up conflicts'" (p. 7). As if the service of science in the defense of democratic ideals had to be sought. The basic strategy of this book was to emphasize the extent of genetic diversity that renders every individual human genetically unique. This meant, among other things, that whatever a "race" is, it is not a uniform genetic type; different races are only populations that differ in the frequencies of some genes. There is, then, too much genetic variation for any genetically substantiated racism (Dunn and Dobzhansky 1946, pp. 90–115). Dunn and Dobzhansky also reasoned, rather cryptically, that "A biologist must assert the absolute uniqueness of every human individual. This same assertion, translated into metaphysical and political terms is fundamental for both ethics and democracy" (p. 46).

In a long article completed in 1946, based on a lecture to the

Princeton American Civilization Program, Dobzhansky made somewhat clearer his view of the relevance of extensive genetic variation to democracy—or more specifically to the American democratic ideal of equality of opportunity. That is, equality of opportunity is the best political system for dealing with extensive genetic diversity because in that system people are allowed to, and are more likely to, assume a role in society for which they are fitted by virtue of their genetically and environmentally determined abilities; in an aristocracy, by contrast, people would not be so likely to find their way into the occupation or social role for which their genes especially fitted them (Dobzhansky 1946, pp. 104–110; see also Dunn and Dobzhansky 1952).

So, far from genetic diversity undermining democracy by resulting in inequality, it serves as a rationale for democratic equality. Later Dobzhansky stated the case even more strongly in *Mankind Evolving*: "Equality of opportunity tends to make the occupational differentiation comport with the genetic polymorphism of the population, and would be meaningless if all people were genetically identical" (Dobzhansky 1962, p. 244).

In the 1940s Dobzhansky was also pursuing a second line of reasoning to minimize the possible inequalities due to genetic diversity. This involved the notion of developmental "plasticity," which Dobzhansky contrasted with genetic "fixity." From a developmental point of view, he argued, evolution by natural selection can lead to two different kinds of adaptations. When the environment is fairly uniform and constant, with only temporary changes from the norm, then there may be one narrow, optimal phenotype. And natural selection may favor developmental pathways that buffer the expression of this phenotype against environmental fluctuations. These would be genetically fixed phenotypes. But when the environment is very variable in space and time, then there may not be any one narrow, optimal phenotype. A more plastic phenotype—that is, the ability to manifest an array of phenotypes corresponding to and triggered by different environments—may be the most adaptive response for every individual.[6]

[6] See, e.g., Dobzhansky 1945; 1946, pp. 148–154; 1956a, pp. 55–85; Dobzhansky and Ashley-Montagu 1947; and Dunn and Dobzhansky 1952, pp. 133–134. With regard to the notions of developmental plasticity, and the related notions of canalization and buffering, Dobzhansky was influenced by Schmalhausen and Waddington. The relationships between the views of Dobzhansky, Schmalhausen, and Waddington on such matters are discussed by Gilbert in this volume.

Dobzhansky reasoned that developmental plasticity must be very important for humans since our environments not only change by themselves (so to speak) but also because we change our own environments so drastically and so often. Given the extremely rapid changes particularly in our social environments, there is an especially high premium on developmentally plastic psychological characteristics. As Dobzhansky and the anthropologist M. F. Ashley-Montagu explained in their influential, coauthored paper "Natural Selection and the Mental Capacities of Mankind" (1947), "Success of the individual in most human societies has depended and continues to depend upon his ability rapidly to evolve behavior patterns which fit him to the kaleidoscope of the conditions he encounters. He is best off if he submits to some, compromises with some, rebels against other, and escapes from still other situations. Individuals who display a relatively greater fixity of response than their fellows suffer under most forms of human society and tend to fall by the way. Suppleness, plasticity, and, most important of all, ability to profit by experience and education are required" (Dobzhansky and Ashley-Montagu 1947, p. 588). "Educability," they argued, was the most important of all developmentally plastic, psychological characteristics of humans.

To the extent that natural selection still operates on humans, Dobzhansky thought that it must operate mostly to promote developmental plasticity and eliminate narrowly fixed mental phenotypes. Dobzhansky and Ashley-Montagu argued that selection for educability was so strong that it had become a universal—transracial—trait.

But Dobzhansky also believed that strong selection for developmentally plastic traits such as educability was fully compatible with considerable genetic diversity underlying those traits. His line of reasoning in this regard is not perfectly clear. The idea seems to be that while developmental plasticity is genetically conditioned, there is no single genotype underlying it—for instance, no single genotype for educability. What is especially important about the compatibility of developmental plasticity and considerable underlying genetic diversity, though, is the consequence with respect to the issue of genetically-based inequalities. That is, as Dobzhansky argued, the broad range of phenotypes that each individual is capable of expressing largely covers up or "eclipses" the genetic variation between individuals (and also between races—Dobzhansky and Ashley Montagu

1947, p. 590). Thus, again, the welfare of human populations or the human species, which depends on large stores of genetic variation, does not depend on great differences in the welfare of individuals.

Developmental plasticity, so important to individual humans, and evolutionary plasticity, so important to populations and species, were related in various ways in Dobzhansky's thinking. For instance, they went hand-in-hand as solutions to a common problem, namely rapidly changing environments. Genetic variation—the basis of evolutionary plasticity—gives human populations and the human species the flexibility to adapt to rapidly changing circumstances. But the world is changing so fast that each individual human also needs to be flexible. Developmental plasticity gives each of us the opportunity to adapt to changing circumstances within our own lifetimes (e.g., Dobzhansky 1956a, pp. 70–71).

Ultimately Dobzhansky was able to draw a much tighter connection between developmental and evolutionary plasticity. This connection had to do with the way in which he came to understand how species maintain the genetic variation that underlies their evolutionary plasticity.

In the early 1940s he began to explain the presence of genetic variation in populations in terms of the higher fitness of heterozygotes over homozygotes. But why would heterozygotes be generally superior? Dobzhansky's views on this matter changed considerably over time (Beatty 1987 and Lewontin 1981). By the mid 1950s he had adopted a position that he owed (to many people, but most directly) to the Russian-American geneticist Michael Lerner, and especially to Lerner's book *Genetic Homeostasis* (1954). The idea was that heterozygotes are superior to homozygotes precisely because they are the basis of developmentally and physiologically more plastic phenotypes. They result in more plastic phenotypes because they consist of two different alleles instead of two copies of the same allele and hence have dual phenotypic capacities relative to the unitary capacities of homozygotes. They are in this way able to cope with a wider variety of environmental conditions. Developmental plasticity is thus part of the causal explanation of the maintenance of genetic variation, which in turn underlies evolutionary plasticity.

By invoking the importance of heterozygote superiority, Dobzhansky was able to show that, in an important sense, the welfare of a population or species and that of its individual members coincide: genetic variation is important at both levels. And yet Dobzhansky's position hardly did away with the individual costs of maintaining variation within a population. Because of the nature of Mendelian inheritance, the superior heterozygotes give rise to some less fit homozygotes in every generation. Thus in his 1955 article "A Review of Some Fundamental Concepts and Problems of Population Genetics," where he defended his new account of the maintenance of genetic variation, he still acknowledged what was by this time a recurrent theme in his work: "Selective processes may at times run into contradictions and, paradoxically, may result in miscarriage of the process of adaptation. Thus, the conflict between the welfare of the individual and of the group to which he belongs is obvious enough in the human species to need no extended comment . . ." (Dobzhansky 1955, p. 11).

Yet another way that Dobzhansky sought to resolve the group/individual conflict—very different from the other approaches I have discussed—involved his and others' attempts in the 1940s and 1950s to articulate more clearly the meaning of the term *fitness*. Dobzhansky was concerned to show that fitness was ultimately a matter of reproductive success and nothing more.[7]

Those who had previously tried to draw moral and political lessons from Darwinian evolutionary theory had been able to do so only by interpreting fitness much more broadly. For instance, in order for early twentieth-century American eugenicists to argue for measures to reduce the high reproductive rates of "obviously" unfit immigrants, they had to identify fitness with something quite different from reproductive success, some presumably inheritable virtues that the older American stock had and that the new immigrants did not.

Diane Paul (1988) has argued that the efforts of Dobzhansky and others to restrict the meaning of the term *fitness* to reproductive success should be seen in part as a reaction to previous political uses, especially eugenic and most especially Nazi eugenic uses. Of course, the change had a theoretical basis as well, but it was enforced at this time by these people for reasons not solely theoretical. This is an important point and represents another instance of the response of American scientists—and geneticists in particular—to the rise of fascism.

[7] See, e.g., Dunn and Dobzhansky 1946; Dobzhansky 1950, pp. 164–165; and Dobzhansky 1956b, pp. 593–594.

But it is worth noting what else was accomplished by Dobzhansky's restriction of the meaning of *fitness* and what else motivated the change. The restriction reduced considerably the individual cost of evolution by natural selection. The less fit individuals do not necessarily suffer for a lack of food or space. The less fit are not necessarily killed or wounded in a struggle for resources. Especially among civilized humans, the less fit are simply those who contribute fewer descendants. As Dobzhansky explained, "At first sight the distinction between the classical and modern variants of the theory of natural selection may seem to be of no great consequence. In reality the difference is important. The classical 'fittest' was the lusty, implacable conqueror in the never-ending struggle for existence with his fellows and with other creatures. Fitness required vigor, power, and ruthlessness. The modern concept emphasizes reproductive success, which means leaving a large number of progeny. Strength and sturdiness are important only insofar as they contribute to this. The fittest is no more spectacular than the parent of the family with the greatest number of surviving members" (Dobzhansky 1956a, pp. 60–61).

And elsewhere, "The 'fittest' is nothing more remarkable than the producer of the greatest number of children and grandchildren. The 'struggle' is usually nothing like physical combat; rather, it is doing one's best to avoid combat and making the best provisions for the welfare of one's family" (Sinnott, Dunn, and Dobzhansky 1958, p. 247).

Maintaining enough genetic diversity to ensure the evolutionary plasticity of a human population or the human species ultimately means only that some individuals will be less fit in the sense of leaving fewer descendants. It is not necessarily any greater cost to the less fit than that—though Dobzhansky was usually quick to add that the costs were often still great (e.g., Dobzhansky 1962, p. 130).

As much as he tried, and in as many different ways as he tried to resolve the group/individual conflict, Dobzhansky never found a solution that completely satisfied him—though he did satisfy himself that the conflict was not so great as had been previously thought, and in particular that the maintenance of genetic variation within human populations or the human species implied no (rational) costs in terms of differential worth.

It is at least some measure of the circulation of Dobzhansky's views that they received considerable attention by American historian Stow Persons in his textbook of American intellectual history, *American Minds* (1958). Persons devoted a long section to the broader implications of recent work in genetics and evolutionary biology. For instance, he noted that a revision of some of the fundamental concepts of Darwinian evolutionary theory had occurred and that "Natural selection thus [i.e., as it is now] conceived is ordinarily a peaceful process." More generally, he reported that "Out of the thinking of scientists on these related problems there emerged in the years after the Great Depression certain principles that were believed to provide a biological basis for democracy" (p. 359). Persons explicitly invoked the authority of Dobzhansky.

CONCLUSION

I have two very different sorts of issues to raise in conclusion. First I want to correct if necessary the understandable impression that Dobzhansky's preoccupation with the group/individual conflict reflects his overwhelming concern for the welfare of individuals. Dobzhansky was clearly concerned. But I believe it is more accurate to say that he was especially interested in the evolutionary welfare of populations and species and that his concern to resolve (to the greatest extent possible) the group/individual conflict was a means of avoiding moral reservations about his pursuits and beliefs. To be more blunt, if not necessarily more precise, Dobzhansky's attempts to resolve the group/individual conflict served in large part to make his pursuits and beliefs more palatable.

He was well aware of the possible effect of moral misgivings upon the reception of particular evolutionary views. There were, moreover, viable alternatives to Dobzhansky's views from which to choose. The empirical evidence alone did not suffice to guarantee the reception of his position. Recall that Muller had argued that natural selection tends to reduce genetic variation. Dobzhansky's first student, Bruce Wallace, once mentioned the unmentionable in an assessment of the relative strengths of Dobzhansky's and Muller's positions: he suggested that Dobzhansky's outlook was "morally inferior" to Muller's since Muller's better guaranteed equality. I believe that Dobzhansky's long-standing concern to resolve the group/individual conflict was in part an attempt to keep such moral problems from detracting too much from his views.

Along the same lines, it is worth recalling that Dobzhansky was a population geneticist, very active in promoting that new discipline. I would suggest that another part of what he was trying to do in raising again and again the apparent conflict between the welfare of the group and the welfare of the individual was to demonstrate the importance of a population perspective. He was concerned to show that population geneticists like himself had something important to add—over and above what geneticists (like Muller) had to say—about evolution. The conflict in question helped to substantiate the irreducibility of population-genetic to genetic issues; one should not simply extrapolate from the deleteriousness of many mutations for their individual carriers to the harmfulness of those mutations for the population.

The second issue I want to raise in concluding has to do with the way in which American scientific response to the rise of fascism is often treated. It is often said that American scientists reacted to the rise of fascism by defending the importance of "disinterested" and politically "autonomous" science. Thus, for instance, in his treatment of the history of nature-nurture disputes in America, Hamilton Cravens argues that "The tumultuous and horrifying events of the thirties and forties persuaded American scientists of the centrality of intellectual liberty and freedom to the free and disinterested pursuit of science, and provided ample evidence of the vulnerability of science to unscrupulous politics" (Cravens 1978, p. 190; see also Ludmerer 1972, pp. 130–132).

Cravens and others are certainly right that this was one of the main agendas of concerned American scientists at the time. But for all their concerns about overly tight connections between politics and science, antifascist American scientists were themselves heavily into political ideology—the ideology of democracy. Intellectual freedom is part of democratic ideology, and perhaps that is why the promotion of democracy looks like no more than the promotion of free inquiry. Thus Cravens continues that, "it was in the spirit of this new awareness of the interdependency of liberal democracy and the pursuit of science that Dobzhansky and Ashley-Montagu wrote their article in *Science* in 1947" (Cravens 1978, p. 290). But intellectual freedom is only a part of democratic ideology. There is a lot left over.

Many biologists of the 1930s, 1940s, and 1950s defended democratic ideals beyond those of intellectual freedom, for example the ideal of American-style egalitarianism. Dobzhansky was one of those. To be sure, his egalitarian agenda promoted his science, but that was perfectly compatible with his sincere attempt to use science to promote democracy.

REFERENCES

Beatty, John. 1987. "Dobzhansky and Drift: Facts, Values, and Chance in Evolutionary Biology." In Lorenz Krüger et al., eds., *The Probabilistic Revolution.* Vol. 2, *Ideas in the Sciences.* Cambridge: MIT Press.

Chetverikov, S. S. 1929? "Mutations in Nature." Unpublished manuscript.

Chetverikov, S. S. 1961. "On Certain Aspects of the Evolutionary Process from the Standpoint of Modern Genetics." Trans. M. Barker, ed. I. M. Lerner. *Proceedings of the American Philosophical Society* 105: 167–195.

Cravens, Hamilton. 1978. *The Triumph of Evolution: American Scientists and the Heredity-Environment Controversy, 1900–1941.* Philadelphia: University of Pennsylvania Press.

Darwin, Charles. 1859. *On the Origin of Species.* London: Murray.

Darwin, Charles. 1871. *The Descent of Man, and Selection in Relation to Sex.* London: Murray.

Dobzhansky, Theodosius. 1937. *Genetics and the Origin of Species.* New York: Columbia University Press.

———. 1938. "The Raw Material of Evolution." *Scientific Monthly* 46: 445–449.

———. 1945. "An Outline of Politico-Genetics." *Science* 102: 234–236.

———. 1946. "The Genetic Nature of Differences Among Men." In Stow Persons, ed., *Evolutionary Thought in America.* New Haven: Yale University Press.

———. 1950. "Heredity, Environment, and Evolution." *Science* 111: 161–166.

———. 1955. "A Review of Some Fundamental Concepts and Problems of Population Genetics." *Cold Spring Harbor Symposia on Quantitative Biology* 20: 1–15.

———. 1956a. *The Biological Basis of Human Freedom.* New York: Columbia University Press.

———. 1956b. "Does Natural Selection Continue to Operate in Modern Mankind?" *American Anthropologist* 58: 591–604.

———. 1962. *Mankind Evolving: The Evolution of the Human Species.* New Haven: Yale University Press.

———. 1962–1963. "The Reminiscences of Theodosius Dobzhansky." Typed transcript. 2 parts. Oral History Research Office, Columbia University, New York.

Dobzhansky, Theodosius, and M. F. Ashley-Montagu. 1947. "Natural Selection and the Mental Capacities of Mankind." *Science* 105: 587–590.

Dunn, L. C., and Theodosius Dobzhansky. 1946. *Heredity, Race, and Society.* New York: Penguin.

————. 1952. *Heredity, Race, and Society,* 2d ed. New York: New American Library.

East, Edward M. 1929. *Heredity and Human Affairs.* New York: Scribners.

Fisher, R. A. 1930. The *Genetical Theory of Natural Selection.* Oxford: Oxford University Press.

Jennings, Herbert Spencer. 1930. *The Biological Basis of Human Nature.* New York: Norton.

Kevles, Daniel J. 1985. *In the Name of Eugenics: Genetics and the Uses of Human Heredity.* New York: Knopf.

Kuznick, Peter J. 1987. *Beyond the Laboratory: Scientists as Political Activists in 1930s America.* Chicago: University of Chicago Press.

Lerner, I. Michael. 1954. *Genetic Homeostasis.* Edinburgh and London: Oliver and Boyd.

Lewontin, Richard C. 1981. "Introduction: The Scientific Work of Theodosius Dobzhansky." In Lewontin et al., eds., *Dobzhansky's Genetics of Natural Populations I–XLIII.* New York: Columbia University Press.

Ludmerer, Kenneth M. 1972. *Genetics and American Society: A Historical Appraisal.* Baltimore: The Johns Hopkins University Press.

Mitman, Gregg. 1988. "From the Population to Society: The Cooperative Metaphors of W. C. Allee and A. E. Emerson." *Journal of the History of Biology* 21: 173–194.

————. 1990. "Evolution as Gospel: William Patten, the Language of Democracy, and the Great War." *Isis* 81: 446–463.

————. Forthcoming. *The Peaceable Kingdom: Animal Ecology and Community at Chicago 1910–1950.* Chicago: University of Chicago Press.

Morgan, Thomas Hunt. 1903. *Evolution and Adaptation.* New York: Macmillan Co.

————. 1932. *The Scientific Basis of Evolution.* New York: Holt.

Muller, H. J. 1929. "The Method of Evolution." *Scientific Monthly* 29: 481–505.

————. 1934. "Radiation Genetics." *Proceedings of the Fourth International Radiologen Kongress,* Zürich, July 1934 2: 100–102.

————. 1950a. "Evidence of the Precision of Genetic Adaptation." *The Harvey Lectures,* series 43, 1947–48 (Springfield, Ill.: Charles C. Thomas), pp. 165–229.

————. 1950b. "Our Load of Mutations." *American Journal of Human Genetics* 2: 111–176.

Paul, Diane B. 1988. "The Selection of the 'Survival of the Fittest'." *Journal of the History of Biology* 21: 411–424.

Persons, Stow. 1958. *American Minds.* New York: Holt.

Popenoe, Paul, and Roswell Hill Johnson. 1923. *Applied Eugenics.* New York: Macmillan.

Purcell, Edward A. 1973. *The Crisis of Democratic Theory: Scientific Naturalism and the Problem of Value.* Lexington, Ken.: University Press of Kentucky.

Rife, David C. 1945. *The Dice of Destiny: An Introduction to Human Heredity and Racial Variations.* Columbus, Ohio: College Book Co.

Sinnott, Edmund W. 1945–1946. "The Biological Basis of Democracy." *The Yale Review* 35: 61–73.

Sinnott, Edmund W., and L. C. Dunn. 1925, 1932, 1939. *Principles of Genetics.* New York: McGraw-Hill.

Sinnott, Edmund W., L. C. Dunn, and Theodosius Dobzhansky. 1950, 1958. *Principles of Genetics.* New York: McGraw-Hill.

Sumner, William Graham. 1934. *Essays of William Graham Sumner,* 2 vols. New Haven: Yale University Press.

Todes, Daniel P. 1987. "Darwin's Malthusian Metaphor and Russian Evolutionary Thought." *Isis* 78: 537–551.

———. 1989. *Darwin Without Malthus: The Struggle for Existence in Russian Evolutionary Thought.* New York: Oxford University Press.

Wright, Sewall. 1931. "Evolution in Mendelian Populations." *Genetics* 16: 97–159.

Dobzhansky in the "Nature-Nurture" Debate

Diane B. Paul

IN "How Much Can We Boost IQ and Scholastic Achievement?" the psychologist Arthur Jensen asserted that genetic differences probably explain at least half of the black-white gap in IQ test scores (Jensen 1969). His article produced a storm of controversy. Initially Jensen was criticized for exaggerating the significance of heritability estimates and for using statistics on the heritability of IQ within races to draw conclusions about the genetics of IQ differences between them. Even his severest critics took for granted, however, that IQ differences within populations were to some degree heritable (Kagan 1969; Hirsch 1970; Lewontin 1970; Bodmer and Cavalli-Sforza 1970).

Within a few years the character of the critique had markedly shifted. Jensen was now faulted, not just for extrapolating from within- to between-group heritability, or presuming that a high heritability implied that environmental measures could do little to change IQ, but for his assumption that genes affect individual differences in intellectual performance. Other behavior geneticists, some quite critical of Jensen, were also indicted for the same offense.

One of the participants in the genetics of IQ debate was Theodosius Dobzhansky. In the 1950s and 1960s, when he published several popular books and articles on the nature-nurture controversy, Dobzhansky had been widely considered an "environmentalist." In the early 1970s, however, he expressed considerable sympathy for Jensen (whose 1972 book, *Educability and Group Differences*, Dobzhansky reviewed favorably in manuscript) and for the besieged community of behavior geneticists in general. He was also angry at some of Jensen's critics. From his letters, we know that Dobzhansky was particularly distressed by these geneticists' refusal to acknowledge that genes made any contribution to individual differences in human cognitive abilities and aptitudes. Indeed, he felt personally offended by their critiques of human behavior genetics.

As well he might. It was a discipline Dobzhansky had done much to foster. He often said that the ultimate aim of any genetics, including his own work with *Drosophila,* was a better understanding of people, especially their behavior. Indeed, he considered the genetics of behavior the next frontier in biology; perhaps "as important and exciting in the near future as molecular genetics has been for our present generation" (1972, p. 523).

In the 1950s and 1960s, Dobzhansky encouraged many young geneticists to enter the field. At the Institute for the Study of Human Variation at Columbia University he provided training and support for behavior genetics research. He served as president of the Behavior Genetics Association and as a member of its Executive Committee. (The association recognizes special achievement in the field with a Dobzhansky Memorial Award). He also convinced some influential social scientists, most notably Gardner Lindzey, then president of the Social Science Research Council, to pay greater heed to genetic determinants of behavior. Dobzhansky himself served on the Research Council's Committee on Genetics and Behavior. There was no basis for human behavior genetics, however, if there was no selective variance for human mental and personality traits.

Dobzhansky believed that human variation in virtually every trait was genetically influenced. "It is a fair statement that whenever a character variable in human populations has been at all adequately studied, genetic as well as environmental components in its variability have been brought to light," he wrote in a typical passage. "This applies to characteristics of all sorts—physical, physiological, and psychological—from skin color, stature, and weight, to intelligence, special abilities, and even to smoking habits" (1973a, p. 283). Dobzhansky made many similar statements—then wholly uncontroversial—in the 1940s, 1950s, and 1960s. However, in the wake of the Jensen controversy, the "interactionist" position Dobzhansky represented came under attack. Within a remarkably brief period, and with his own views having shifted little if at all, Dobzhansky had apparently moved from one side of the nature-nurture debate to the other.

What had changed, with amazing rapidity, was the social context of behavior genetic research. In the politically charged 1970s, the nature-nurture controversy had been transformed, and with it the meanings of key concepts and terms. Views that in the 1960s were considered "environmentalist" now marked one as "hereditarian." And "hereditarian" in the 1970s implied politically reactionary. Dobzhansky (like many of his colleagues) was bewildered by this reversal. He did not consider himself a hereditarian. An active participant in many left-liberal causes, above all antiracialism, he certainly did not consider himself a reactionary.

In this essay I explore Dobzhansky's own (sometimes shifting and contradictory) views on the nature-nurture issue and the allied question of eugenics. To what extent did he think individual and group differences in mentality and behavior were attributable to differences in genes? Did he believe the answer mattered for questions of social policy? If so, in what ways? Above all, how did circumstances change, such that a vigorous and consistent critic of genetic determinism came, at least partially, to sympathize with some of those tagged as hereditarians?

Following World War II Dobzhansky wrote frequently about genetics and society in general and the nature-nurture question in particular. Most of these works, such as *Heredity, Race, and Society* (Dunn and Dobzhansky 1946), *Mankind Evolving* (1962), *Heredity and the Nature of Man* (1964), and *Genetic Diversity and Human Equality* (1973b), addressed a popular audience. In these books and in a host of articles, Dobzhansky portrayed himself as a moderate, rejecting the "extremes" of environmentalism (defined as the view that genes are irrelevant to mentality and behavior) and hereditarianism (defined as the view that only genes matter). The following passage is typical: "It would please many simplifiers to have the diversity of human abilities and behaviors either all due to training or all predetermined genetically. The partisans of these two oversimplifications engage in interminable polemics . . ." (1972, p. 529).

In fact, these positions have had few exponents in America, at least among geneticists. Even extreme eugenicists have generally accorded some role to the environment. And, excepting a brief period in the 1970s, few environmentalists have insisted that genes contribute nothing to mental and personality differences. For most of the history of the nature-nurture debate, the real differences have been ones of degree and of perceived social implications.

In general, environmentalists have accorded greater weight to nurture in explaining human differences, have challenged the equation of "genetic" with "fixed," and have denied that genetic knowledge justifies prevailing social arrangements. Hereditarians, on the contrary, have given greater weight to genes, which are seen as determining both individual fates and social institutions.

In his 1949 book *The Nature-Nurture Controversy*, Nicholas Pastore described the positions as they existed about the time the subject first captured Dobzhansky's attention—descriptions that remained applicable throughout the 1950s and 1960s. According to Pastore, a hereditarian "accepts statements of the following type: heredity is more important than environment; individual and group differences are the result of innate factors (either in totality or predominantly); innate characteristics are not easily modified. Where a choice of interpretation is possible, the explanation in genetic terms is the one advanced and favored. To the hereditarian way of thinking, the problem of differential fecundity looms as a most significant one for society" (p. 14). An environmentalist, by contrast, "accepts statements of the following type: environment is more important than heredity; existing individual and group differences reflect (much more than is commonly thought) differences in opportunity; innate characteristics are easily modified. Furthermore, the 'plasticity' of the child is emphasized. Of possible alternative explanations, he chooses the one emphasizing environment. In addition, the environmentalist minimizes the importance of natural inequities in the attainment of success and rejects the eugenic program (as usually conceived)" (p. 14).

If Dobzhansky rejected the label, he nonetheless held generally "environmentalist" views. He insisted that we are not blank pages on which the environment writes, but he also insisted on the significance of social and cultural differences. While believing that populations, as well as individuals, vary genetically in respect to many traits, he valued these genetic differences and wished to preserve rather than reduce them. ("Differences are not deficits" was one of his favorite maxims.) He was thus generally critical of eugenics. He argued strenuously against the use of genetics to justify existing class structures or racial prejudice. Above all, he did more than anyone to undermine the assumption that heritability can be equated with insensitivity to environmental change.

Dobzhansky took every opportunity to note that the same genotype may be expressed differently in different environments. Human environments are both diverse and everchanging. "Invention of a new drug, a new diet, a new type of housing, a new educational system, a new political regime introduces new environments" (1955, p. 75). Thus we can have at best only incomplete knowledge of the "norm of reaction" of any human genotype. If we cannot identify the range of environments over which genotypic expression varies, we cannot predict how much the intellectual performance (for example) of an individual or population might rise in another environment.

Estimates of heritability (the proportion of phenotypic variance attributable to genetic variance) thus provide local rather than global information; they apply only to a specific population in a specific range of environments. "Heredity is often spoken of as 'destiny'," he wrote. "It is destiny largely in proportion to our biological ignorance" (1950, p. 162).

Yet Dobzhansky sometimes wrote in quite a different vein. He favored a meritocracy, in which the occupational differentiation of the population would comport with its genetic polymorphism. "Civilization fosters a multitude of employments and functions to be filled and served—statesmen and butchers, engineers and policemen, scientists and refuse collectors, musicians and sales clerks," he wrote in *Mankind Evolving* (1962, p. 243). The ability and desire to do these jobs, in his view, had a strong genetic component. "A society benefits from the fullest development of genetically conditioned and socially useful talents and abilities of its members" (1973a, p. 283); it should capitalize on genetic diversity in the service of social efficiency.

In his comments on the natural division of labor, Dobzhansky tended to write as if environments were static and heredity were indeed destiny. However, when discussing the significance of the norm of reaction concept, he emphasized the difficulty of knowing who was best at what, given that the same genotype may be expressed differently in different environments. Thus he wrote, "in different environments and under different social systems the present failures might be successes and the successes failures" (1962, p. 314). As Michael Ruse notes in his essay in this volume, Dobzhansky's social attitudes were complex and sometimes contradictory. In this case he appears torn between commitments to the values of social efficiency and social experimentation.

Dobzhansky's surge of writing on these issues seems to have been prompted by his association with Frederick Osborn, the director of the American Eugenics Society. Following World War II, Osborn embarked on an effort to repair the society's image, which had been badly tarnished by its association with crude class and race prejudice. Osborn aimed to place eugenics on a firmer scientific foundation and, to this end, tried attract distinguished scientists to the organization. Among those he courted most assiduously was Dobzhansky, who finally joined the society's Board of Directors in 1964. By then Dobzhansky and Osborn had become quite close. During the 1950s and 1960s, Osborn's views became increasingly

"reformist"; i.e., he came more and more to emphasize the importance of environment in explaining both individual and group
differences and to move the society away from propaganda for action toward support for scientific research. Osborn's move in this direction seems to have been principally a result of his association
with Dobzhansky.

At the same time, Dobzhansky became embroiled in a dispute
with H. J. Muller over the nature and consequences of genetic diversity—a debate that came to be known as the "classical-balance" controversy. Their quarrel began with Muller's 1949 presidential address
to the American Society of Human Genetics, published the following
year as "Our Load of Mutations." The essay was extremely influential. *Genetic load* and various associated terms came to dominate the
discourse of population genetics. This is not the place for a detailed
explanation of these concepts, which have been discussed by John
Beatty in this volume and elsewhere. Suffice it to say that *genetic
load* is not a neutral expression; it implies that variation is a burden
that we carry, i.e., that it is bad.

This attitude toward genetic variability followed logically from
Muller's view that organisms are generally extremely well adapted to
their environments. Thus nearly all mutations are bad and will be
removed by selection. Of course favorable mutants sometimes appear, and these provide the raw material for evolution. But they are
extremely rare and rapidly become the new normal or "wild type."
Most genetic variation is therefore transitory. More accurately, it
would be transient in nature. However, humans have both increased
the rate of mutation (through increased exposure to mutagens, especially ionizing radiation) and decreased that of selection (primarily
through improvements in medicine and public health). As a result,
the species is genetically deteriorating. Given these assumptions, the
social policy implications seemed clear: there was an urgent need for
a eugenics program.

Dobzhansky was perhaps the severest and certainly the most influential critic of Muller's eugenics and of its underlying assumptions. In Dobzhansky's view also, some variation was unreservedly
bad. But he stressed the heterogeneous and changing character of
environments and hence the need for a store of genetic variability.
Given this need, selection would generally act to preserve variation.
As he wrote in a letter dated 11 July 1953 to Julian Huxley: "It does look
that balanced polymorphism is of greater importance in adaptive

evolution ... than we have imagined. ... This may mean that what we regarded as lethals and hereditary diseases are in reality the raw materials from which the species constructs the co-adapted gene combinations. It will be very useful to consider from this standpoint some of the old problems of human genetics—and eugenics, of course" (Huxley correspondence, Rice University). In other words, disability and disease may be the price a species pays for evolutionary adaptability.

By the late 1950s Dobzhansky had come to focus on a single form of balancing selection: heterozygote advantage, or "overdominance." If heterozygotes are generally fitter than homozygotes, then genetic variability is good for individuals as well as species. Dobzhansky's student Bruce Wallace had irradiated fruit flies and found that the treated group, with their induced heterozygosity, had a greater viability than the controls. These experiments were sometimes cited as evidence for the virtue of heterosis or heterozygosity per se. Heterosis explained why some harmful genes were maintained at high frequency in the population; for example, the allele that in double dose produces a serious disease, sickle-cell anemia, produces, when paired with a normal allele, only mild effects and a more than offsetting protection against malaria. One could not, and would not want, to select against genes of this type. If overdominance were common, eugenics would thus be futile. (Muller conceded the sickle-cell example but denied its generality.)

In Dobzhansky's view, Muller seriously underestimated the value of diversity, both genetic and social. Muller's eugenics was misguided (according to Dobzhansky) because it aimed at an evolutionarily disastrous uniform type. It was really Muller's positive eugenics, rather than his negative eugenics, that roused Dobzhansky's ire. His attitude toward the latter was actually quite complex and perhaps contradictory.

Dobzhansky often wrote as if genetic defects were the necessary price the species pays for evolutionary flexibility. However, he also sometimes wrote that carriers of serious diseases should be convinced—and failing that compelled—not to reproduce. Thus, in *Mankind Evolving* he suggested that: "Persons known to carry serious hereditary defects ought to be educated to realize the significance of this fact, if they are likely to be persuaded to refrain from reproducing their kind. Or, if they are not mentally competent to reach a decision, their segregation or sterilization is justified. We

need not accept a Brave New World to introduce this much of eugenics" (1962, p. 333).

Dobzhansky was interviewed for Columbia University's Oral History Project a few months after the book appeared. Asked to elaborate on this passage from *Mankind Evolving*, he discussed the case of
retinoblastomic children. If you save their lives, he said, in each generation you would create more lives of the same kind to be saved.
Thus, those saved ought to be prevented from reproducing (1962–
1963, p. 443). In cases of serious disease, Dobzhansky thought that all
reasonable people would agree with this prescription. But consensus
would not be reached in respect to more common ailments, like dia-
betes. And as regards which traits actively to foster, consensus would
completely break down. Thus his ire was aroused much more by
Muller's schemes to improve humankind mentally and morally than
by his negative eugenics, which aimed at reducing disease.

In his 1935 tract *Out of the Night*, Muller had proposed mass artificial insemination of women with the sperm of men superior in intel-
lect and character. "In the course of a paltry century or two," he pre-
dicted, "it would be possible for the majority of the population to
become of the innate quality of such men as Lenin, Newton, Leo-
nardo, Pasteur, Beethoven, Omar Khayyam, Pushkin, Sun Yat Sen,
Marx . . . or even to possess their varied faculties combined" (p. 113).
Dobzhansky was horrified by this proposal. In *Mankind Evolving*, he
charged (with a certain degree of exaggeration) that the logical ex-
tension of Muller's philosophy would be selection of "the ideal man,
or the ideal woman, and to have the entire population of the world,
the whole of mankind, carry this ideal genotype" (p. 329). He thought
that prospect disastrous not only from a genetic but also from a social perspective.

Dobzhansky prized human social and cultural differences. And he
saw these as linked to—indeed dependent on—genetic diversity. As
we have seen, Dobzhansky believed as strongly as Muller that
differences in intelligence and temperament are genetically influenced. Indeed, he sometimes expressed stronger views, as in his (oft-
repeated) comment that equality of opportunity "would be meaningless if all people were genetically identical" (1962, p. 244). (Such a
statement would itself be meaningless unless all variation were genetic). The real argument between Dobzhansky and Muller was not
over the existence but over the value of these differences. Dobzhansky always insisted that "Genetic diversity is a blessing, not a curse"
(1971, p. 23). Muller did not really wish to create a population of

human clones. But he would certainly have welcomed a significant increase in the population average for intelligence and various skills and personality traits. When Dobzhansky asked "Do we really want to live in a world with millions of Einsteins, Pasteurs, and Lenins?" the question was rhetorical, for the answer was self-evidently in the negative (1962, p. 330). He was willing to eliminate genes for clear-cut diseases. But he would not interfere when it came to psychic traits. Dobzhansky sometimes wrote that "Eugenics will eventually come into its own" (1973b, p. 49). What (if anything) he meant by that is never clearly expressed. In respect to their cognitive abilities and temperaments at least, Dobzhansky seemed quite satisfied with people the way they were.

That attitude was at least in part a reflection of the view that all kinds of people were needed to do the world's work. As noted earlier, Dobzhansky believed that "Any human society, from the most primitive to the most complex (the latter more than the former), needs a diversity of men adapted and trained for a diversity of functions" (1973b, p. 44; see also 1962, p. 243). In his view, it was unlikely that the requisite diversity could be achieved simply by differential training. Most people could certainly do most jobs—but not all of them. Dobzhansky believed that he himself could have been brought up to be a peasant, a clerk, an engineer, or a soldier, but not a concert pianist, painter, boxer, sprinter, or mathematical prodigy. "Educability is not limitless," he wrote (1973a, p. 287). Moreover, the fact that one can do a job does not mean that one does it easily or with pleasure. Thus the ideal is a society in which all choose the occupation for which they are "most qualified genetically" (1973a, p. 284).

To make the most of genetic diversity, society must equalize opportunities. Only with uniform environments will differences in genetic merit be manifest. The result will be a genetic meritocracy. "With anything approaching full equality, every trade, craft, occupation, and profession will concentrate within itself those who are genetically most fit for those roles" (1973b, p. 45). Dobzhansky stressed that equalizing environments would enhance genetic differences among both individuals and groups since the greater the equality of opportunity in a society, the more the differences among its members are likely to reflect genetic differences (1962, p. 247; 1973b, pp. 29, 33).

However, Dobzhansky's social views were considerably more radical than his comments on meritocracy might suggest. In particular, he argued that the link between occupational status and financial reward should be broken. If there were to be any economic inequalities, they should favor those who do the dirty and dangerous tasks, not the high-status, pleasant ones. "Manual labor is not intrinsically inferior to intellectual labor," he wrote, "even though more people may be adept at the former than at the latter" (1973b, p. 49), and he explicitly endorsed Marx's dictum "From each according to his abilities, to each according to his needs" (1973b, p. 42). Elsewhere he quoted approvingly the sociologist Christopher Jenck's comment that progress toward the goal of economic equality requires establishing political control over our economic institutions and that "this is what other countries call socialism" (1973a, p. 288). However, Dobzhansky's view that history was moving inexorably in the direction of greater economic equality was perhaps rather unrealistic.

In any case, until the 1970s Dobzhansky's views on the nature-nurture question would certainly have marked him as an "environmentalist." Consider in this respect Pastore's analysis (1949) of the political correlates of attitudes to the nature-nurture question. His book consists of twenty-four profiles of scientists active in the debate; twelve are characterized as hereditarians, twelve as environmentalists. H. S. Jennings could have spoken for most of Pastore's environmentalists when he wrote: "It is certain that all the things that affect character and conduct are deeply influenced by the hereditary materials. There is no characteristic or quality that is exempted from its influence. This conclusion is confirmed by all the many studies that have been made on the two types of twins. And it is in harmony too with all that we know of the science of genetics" (Jennings 1935, p. 204).

In 1949 few would have quarreled with Pastore's assignments to either category. Through the 1950s and 1960s, the positions remained stable. Among geneticists, Jerry Hirsch would emerge as one of Arthur Jensen's severest critics. But in the 1960s Hirsch did not dispute the heritability of human mental traits. Indeed, only two years prior to Jensen's article he wrote: "As the social, ethnic, and economic barriers to education are lowered throughout the world and as the quality of education approaches a more uniformly high level of effectiveness, heredity may be expected to make an ever larger contribution to individual differences in intellectual functioning and consequently to success in our increasingly complex civilization" (1967, pp. 434–435). As noted earlier, the first scientific critiques of

Jensen's article assume a fairly high heritability of intelligence; they object principally to the import accorded heritability estimates and the extrapolation from individuals to races.

But in the highly charged atmosphere of the early 1970s, the critique of Jensen began to broaden. This shift in approach was spurred by the scandal involving the work of British psychometrician Cyril Burt. In a 1974 book, *The Science and Politics of IQ*, Leon Kamin charged that Burt's influential results (which apparently demonstrated an 80+ heritability of IQ) were, statistically speaking, too good to be true. This is not the place to review the history of the scandal that followed. Suffice it to say that Kamin's suspicions were justified; Burt had apparently fabricated at least some of his work. The ensuing scandal led to a reanalysis of other classic studies of the heritability of IQ. All were judged and found wanting by contemporary methodological standards. In light of these critiques—which focused on studies purportedly demonstrating a high heritability of IQ within the white population—some began to question conventional assumptions concerning individual differences in intelligence and personality.

Within just a few years, the critical position had been transformed. For example, after reviewing all the classic studies of the heritability of IQ, Kamin concluded: "There exist no data which should lead a prudent man to accept the hypothesis the IQ test scores are in any degree heritable" (1974, p. 1). Soon, critics of Burt, Jensen, and Richard Herrnstein (who advanced an argument in respect to social class analogous to Jensen's concerning race; 1971, 1973) appeared to commit themselves to the proposition of zero heritability of any interesting skill or behavior. That was obviously not a perspective that Dobzhansky could share. Indeed, it denied assumptions he considered self-evidently true.

The debate of the 1970s focused on two issues: whether it was possible to design human behavior genetic studies that met reasonable methodological standards; and whether the effort to do so was justified by their potential scientific or social value. For most critics, the answer to the first was "probably," and to the second, a certain "no." For Dobzhansky, the answer to both was a certain "yes." That is why, in spite of a position remarkable for its consistency over a thirty-year period, he came to find himself first on one side and then the other of the nature-nurture debate. He was both baffled by and indignant at this development. Had he lived another five years, however, he would have seen the debate shift once again—virtually back to where it was in the 1960s.

There remain important issues in dispute. But for the most part, they are disputes among interactionists. Moreover, the point now stressed by critics of contemporary behavior genetics is in fact the point Dobzhansky himself did most to popularize (even if he did not always acknowledge its full implications): Heritability estimates apply only to a specific population in a specific range of environments. Unless we know the full range of environments over which genotypic expression may vary, we are not justified in assuming the ineffectiveness of environmental change. And that is information we do not, and indeed cannot, possess.

References

Bodmer, W. F., and L. L. Cavalli-Sforza. 1970. "Intelligence and Race." *Scientific American* 223: 19–29.

Dobzhansky, T. 1950. "Heredity, Environment, and Evolution." *Science* 111: 161–166.

———. 1955. *Evolution, Genetics, and Man.* New York: John Wiley and Sons.

———. 1962. *Mankind Evolving: The Evolution of the Human Species.* New Haven: Yale University Press.

———. 1962–1963. "The Reminiscences of Theodosius Dobzhansky." Typed transcript. 2 parts. Oral History Research Office, Columbia University, New York.

———. 1964. *Heredity and the Nature of Man.* New York: Harcourt, Brace and World.

———. 1971. "Race Equality." In R. Osborne, ed., *The Biological and Social Meaning of Race* (San Francisco: W. H. Freeman), pp. 13–24.

———. 1972. "Genetics and the Diversity of Behavior." *American Psychologist* 27: 523–530.

———. 1973a. "Is Genetic Diversity Compatible With Human Equality?" *Social Biology* 20: 280–288.

———. 1973b. *Genetic Diversity and Human Equality* (New York: Basic Books).

Dunn, L. C., and Th. Dobzhansky. 1946. *Heredity, Race, and Society.* New York: New American Library.

Herrnstein, R. 1971. "IQ." *Atlantic Monthly* 228: 43–64.

———. 1973. *IQ in the Meritocracy.* Boston: Little, Brown.

Hirsch, J., ed. 1967. *Behavior-Genetic Analysis.* New York: McGrawHill.

———. 1970. "Behavior-Genetic Analysis and Its Biosocial Consequences." *Seminars in Psychiatry* 2: 89–105.

Jennings, H. S. 1935. *Genetics*. New York: Norton.

Jensen, A. 1969. "How Much Can We Boost IQ and Scholastic Achievement?" *Harvard Educational Review* 39: 1–123.

Kagan, J. S. 1969. "Inadequate Evidence and Illogical Conclusions." *Harvard Educational Review* 39: 274–277.

Kamin, L. 1974. *The Science and Politics of IQ*. Potomac, Maryland: Lawrence Erlbaum.

Lewontin, R. C. 1970. "Race and Intelligence." *Bulletin of the Atomic Scientists* 26: 2–8.

Muller, H. J. 1935. *Out of the Night: A Biologist's View of the Future*. New York: Vanguard Press.

———. 1950. "Our Load of Mutations." *American Journal of Human Genetics* 2: 111–176.

Pastore, N. 1949. *The Nature-Nurture Controversy*. New York: Kings Crown Press.

Dobzhansky and the Problem of Progress

Michael Ruse

I START WITH a paradox. Pierre Teilhard de Chardin was a French Jesuit, also a paleontologist, who created a remarkable world picture synthesizing science and religion, in which everything pointed progressively to a supposed "Omega Point," something Teilhard identified with Jesus Christ (Teilhard de Chardin 1955). Expectedly, conventional scientists were scathing in their negative reactions (e.g., Medawar 1969). And yet he who has good claim to having been the greatest evolutionist since Charles Darwin, Theodosius Dobzhansky, was president of the American branch of the Teilhard Society (see Dobzhansky 1967). Unraveling this paradox takes us to the heart of Dobzhansky's thought, including his science.

WAS DOBZHANSKY A PROGRESSIONIST?

I shall argue that the key to the paradox lies in the notion of progress, so let me start right in with this. Was Dobzhansky a progressionist and, if so, of what kind? Answering this question presupposes the answer to a previous question, namely precisely what one means by "progress." Drawing on the work of others, I will assume that progress involves a sequence of events or phenomena, ordered from beginning to end, which may or may not include temporary reversals, which may or may not be completed, but which must in some sense incorporate an increase in value. Mere change is not enough. In some way, things must be getting better (Ayala 1974). The values could be comparative, but here I am more interested in absolute notions. Nevertheless, the choice of values still depends on one's interests—in the social or cultural realm they might involve (say) increased material prosperity, decreased poverty and ignorance, greater acceptance of democracy (Almond et al. 1982).

We should note here that there is a distinctive subclass of notions which have been applied to biological evolutionary change. Popular proposals include increased intelligence and greater complexity

(Nitecki 1988). Invariably, people have humanlike features in mind when they are defining biological progress, but they usually want to avoid blatant circularity. Hence, some independent criterion is sought—a criterion which, it just so happens, we humans possess abundantly. (Not necessarily more abundantly than every other organism, potential or actual. Your Englishman might lament that we are not as loving as the dog.)

Now against this prolegomenon, let us return to our question about Dobzhansky and progress, and since obviously the reason why we are interested in Dobzhansky is because he was such an important biologist, let us ask specifically about Dobzhansky and biological progress. At a general level, the answer is easy to find. Dobzhansky was a progressionist, and, qua biologist, he falls into the usual pattern in thinking that the value-culmination of evolution is humankind.

> The universe, inanimate as well as animate matter, human bodily frame as well as man's psyche, the structure of human societies and man's ideas—all have had a history and all are in the process of change at present. Moreover, the changes so far have been on the whole, though not always progressive, tending toward what we men regard as betterment. Progress in the future is not inevitable; it is not vouchsafed by any law of nature; but it may be striven for. (Dobzhansky 1962, p. 1)

And elsewhere he states:

> Judged by any reasonable criteria, man represents the highest, most progressive, and the most successful product of organic evolution. The really strange thing is that so obvious an appraisal has been over and over again challenged by some biologists. Suppose, it has been argued, that evolution is studied not by man but by a fish. Would not the highest form of animal then have to be a fish? To which Simpson has replied: "I suspect that the fish's reaction would be, instead, to marvel that there are men who question that man is the highest animal. It is not beside the point to add that the 'fish' that made such judgments would have to be a man!" (Dobzhansky 1956, p. 86)

But now let us go on to ask about the particular criterion of (biological) progress that Dobzhansky himself favors. What is it in evolution that is of value, that increases in extent, and that is most fully manifested in our own species? For us the answer is in some sense bound up with our culture, but this is a specific instance of flexibility or adaptability.

Less than two thousand years ago, the ancestors of most modern Americans and Europeans were barbarians eking out a rough and precarious existence in the forests and swamps of northern Europe. But these barbarians responded magnificently when they were given an opportunity to borrow foreign cultures developed by the rather different peoples who inhabited the lands around the eastern part of the Mediterranean Sea. Successful cultural reconditioning can now be observed any day in the large universities in such centers as New York, London, Paris, or Moscow, which attract students from all over the world. At least some of these students become culturally more similar to each other and to their hosts than they are to some of their own biological relatives who have stayed at home. Evidence that cultural patterns are determined by the genes is utterly lacking. What evidence there is rather contradicts this hypothesis. The greatest fallacy of the racist hypothesis is, then, its failure to take into account the adaptive importance of the behavioral plasticity in man. (Dobzhansky 1956, p. 47)

Note that for all the significance of culture, this is a biological progress:

The transition from the adaptive zone of a prehuman primate to the human adaptive zone was brought about by the development of the biological basis for the ability to use symbolic thought, language, to profit by experience, to learn, in short by the development of educability. (Dobzhansky 1956, p. 121)

It is important to note that, for Dobzhansky, this adaptability was certainly not something confined exclusively to humankind, even though we may be more talented in this respect than any other species. More significantly, it was not something confined exclusively to the individual. In those happy presociobiological days when Dobzhansky was writing, when genes were allowed to be as altruistic toward their fellows as Mother Teresa, Dobzhansky, like virtually everyone else, moved happily from the level of the individual to the level of the group. Indeed, in his classic *Genetics and the Origin of Species* (1937), he even has a forerunner of the "species-as-individuals" thesis, for he argues that there are good reasons for regarding the species as a kind of supraindividual (see Hull 1976). With respect to adaptability, it was central to Dobzhansky's biological thinking that groups, the better groups at least, have lots of internal genetic variation and thus are able to respond at once to life's challenges, without need to wait on the arrival of fortuitous variation (mutation).

... the accumulation of germinal changes in the population genotypes is, in the long run, a necessity if the species is to preserve its evolutionary plasticity. The process of adaptation can be understood only as a continuous series of conflicts between the organism and its environment. The environment is in a state of constant flux, and its changes, whether slow or catastrophic, make the genotypes of the past generations no longer fit for survival. The ensuing contradiction can be resolved either through the extinction of the species, or through a genotypical reorganization. A genotypical change means, however, the occurrence of a mutation or of mutations. But nature has not been kind enough to endow the organism with ability to react purposefully to the needs of the changing environment by producing only beneficial mutations where and when needed. Mutations are random changes. Hence the necessity for the species to possess at all times a store of concealed, potential, variability. This store will presumably contain variants which may never be realized in practice, and still other variants which were neutral or harmful at the time when they were produced but which will prove useful later on. (Dobzhansky 1937, pp. 126–127)

Parenthetically, I note that in thus asking a crude question about Dobzhansky and progress, I am thereby collapsing into one the work of a lifetime, assuming that there was no change (progress?) between Dobzhansky's early views (say in the 1930s, or even earlier during the Russian period) and his later views (in the 1960s, when he was writing extensively on our own species). At the level I am working here, I think this crudity is excusable, but I am sure there were some changes. We know, for instance, that Dobzhansky's thinking about the causes of the maintenance of group internal genetic variation certainly changed (Beatty 1987a; Provine 1986).

One final point and then we can move on. Nothing in this world comes easily. There is a cost to adaptability. At both individual and group level, you are never as well-off as you would be if you were perfectly adapted to the present situation. There is a downside, both biologically and (for us) in human terms. Let me quote from the passage given immediately before the one given just above.

It is not an easy matter to evaluate the significance of the accumulation of germinal changes in the population genotypes. Judged superficially, a progressive saturation of the germ plasm of a species with mutant genes a majority of which are deleterious in their effects is a destructive process, a sort of deterioration of the genotype which threatens the very existence of the species and can finally lead only to its extinction. (Dobzhansky 1937, p. 126)

Indeed, Dobzhansky went so far as to say that there will probably be a cost even to standing still (because of such mechanisms as balanced heterozygote fitness). But the really heavy charge comes when we want flexibility for the future.

WHY DID DOBZHANSKY BELIEVE IN BIOLOGICAL PROGRESS?

It may have been "obvious" to Dobzhansky that there is biological progress, but (as he himself acknowledged) this was certainly not something that is obvious to all his fellow biologists. Why, then, did he himself believe in such progress? One can give several reasons, and I am inclined to think that there is merit in all, although one is certainly more important than the others.

First, there is Dobzhansky's Russian background. Thanks to our increased sensitivity to this point, it is no longer necessary to belabor the fact that Dobzhansky was not merely an American with an odd accent (Lewontin et al. 1981). His country of origin may have been materially poor, but—especially with respect to evolutionary studies—it was culturally rich. And of particular pertinence to us here is the fact that Russian evolutionary biology was unequivocally progressive (Adams 1980a). From its Germanic origins to the giants of the twentieth century—A. N. Severtsov (1929) and I. I. Schmalhausen (1949)—a belief in an upward increase in value has been the backbone of Russian evolutionary thought (Adams 1980b). Had Dobzhansky not been a progressionist—at least, had Dobzhansky not started as a progressionist—there would have been cause for comment. Interestingly, in the 1940s it was Dobzhansky who was responsible for the translation into English of Schmalhausen's major work on (progressive) evolution.

Second, there is Dobzhansky's scientism. By this I mean his belief in the worth of science, and the connection I am implying is one that regards science itself as something progressive (probably toward an increasing correspondence with the truth) and which at some level transfers this progressiveness analogically to other domains, including the biological. I am sure this is how things worked for some other evolutionists, most notably J. B. S. Haldane. Haldane went at everything in a progressivist manner because of his almost religious attitude to the power of science (Haldane 1924).

But this was also true to some extent for Dobzhansky. He was certainly committed to a belief in the worth and progressive nature of science. In later life, especially under the influence of his student Francisco Ayala, Dobzhansky thought of himself as a Popperian ((Ayala and Dobzhansky 1974); but the underlying view predated this.

> Science is cumulative knowledge. This makes scientific theories relatively impermanent, especially during the epochs when knowledge piles up in something like geometric progression. Scientists should be conscious of the provisional and transient nature of their attainments. Any scientist worthy of his salt labors to bring about the obsolescence of his own work. (Dobzhansky 1962, p. xii)

And there are many places where Dobzhansky showed that he regarded the cultural and the biological as essentially one.

> The fact which must be stressed, because it has frequently been missed or misrepresented, is that the biological and cultural evolutions are parts of the same natural process. This process, human evolution, must eventually be brought under human control. Here mankind will meet the greatest challenge of its biological and cultural histories. To deal with this challenge successfully, knowledge and understanding of evolution in general, and of the unique aspects of human evolution in particular, are essential. (Dobzhansky 1962, p. 22)

Third, there is the uncomfortable question of eugenics. Massive evidence is now starting to come in showing just how important eugenical beliefs were in the genesis of the science of genetics (Adams 1990). At the populational level, the classic case was R. A. Fisher (1930), who basically created population genetics in order to underpin his hereditarian thoughts about society. But Fisher, although extreme, was not unique. Many (almost all) of the evolutionary biologists of the early days of the Darwin-Mendel synthesis believed in eugenics, and this was tied to beliefs about biological progress. It was thought that we have a moral obligation to keep up the human race, and on the one side this was linked to specific claims about the superiority of certain humans (white, healthy, intelligent) over other humans (black, diseased, stupid), and on the other side this was linked to general claims about the superiority of certain organisms (mammalian, primate, humanoid) over other organisms (invertebrate, cold-blooded, nonviviparous).

In the case of Dobzhansky, the relationship is somewhat tangled. He hated views that smelled in any way of a master race, and he was adamant always that every human being has a unique existential worth. Society needs its hewers of wood and drawers of water as much as, if not more than, its philosophers and fruit-fly geneticists (Dobzhansky 1962). In the 1950s he was locked in bitter battle with fellow geneticist H. J. Muller over the claim that there is an ideal type for the human species (Muller *pro*, Dobzhansky *con*; for details, see Beatty 1987b).

However, there are passages—passages that occur cheek-to-jowl with affirmations of human equality—that show not only that Dobzhansky did not believe in biological identity but that at some level he valued some (genetic) types over others.

> People who should be able to provide the best environment for the physical and mental development of their children produce fewest progeny. Genetic consequences cannot, however, be ignored. They have been debated in many ways by many biologists, psychologists, sociologists, and political propagandists. Many dreadful prophecies and strident proclamations have been made. It cannot be gainsaid that there is a predicament here which should cause concern. (Dobzhansky 1962, p. 313)

A passage like this sounds like Fisher in full flight. Hence, I certainly would not want to claim that eugenics was entirely irrelevant to Dobzhansky's views on progress.

I come now to my fourth and final reason for Dobzhansky's commitment to progress, specifically biological progress but actually all types of progress. This centers on Dobzhansky's religious views, and that I not seem coy, let me state unambiguously that I regard this as the overriding reason for Dobzhansky's faith in progress. Everyone who knew him has testified to the strength of Dobzhansky's religious beliefs. Fellow synthetic theorist G. L. Stebbins spoke for everybody in his assessment that "I would say that of all the scientists I have known and admired, Doby came closest to being a really religious man" (Interview, May 1988). Formally Dobzhansky was Russian Orthodox, but he admitted that this was more a traditional stance, and doctrinally he seems not to have been locked in one way or the other.

> Modern man must raise his sights above the simple biological joys of survival and procreation. He needs nothing less than a religious synthesis. This synthesis cannot be simply a revival of any one of the existing religions, and it may not be a new religion. The synthesis may be

grounded in one of the world's great religions, or in all of them together. My upbringing and education make me biased in favor of Christianity as the framework of the synthesis. I can, however, understand
people who would prefer a different framework. What is important is that
the outcome must be truly a *synthesis*. (Dobzhansky 1962–1963, p. 56)

His ecumenical approach to religion did not imply indifference.
Dobzhansky's faith in God was strong, and his hope—his rather
desperate hope—of salvation and everlasting life was nigh overwhelming.

It is from within Dobzhansky's lifelong Christian faith that we find
his strong acceptance of progress (of all kinds), and it is because of
this Christianity that Dobzhansky reacted so warmly to Teilhard,
whose progressionism, rooted in a metaphysical synthesis of science
and religion, spoke directly to Dobzhansky. By his own admission,
Dobzhansky took up evolutionary studies in the first place (in Russia, before the Revolution) because of his concerns about humankind, and then and forever more he enthused about evolution because its perceived progressionism spoke directly to his Christian
beliefs.

Evolution is the part of biology which has the highest, the most direct
implications, the most reflections in fields not immediately connected
with biology—sociology, philosophy. . . .

Now, this sociological-philosophical angle was really the aspect
which interested me most in the whole field of biology, from the earliest
days, I think really from my first reading of Darwin at about age fifteen.
It is hardly surprising that both during the pre-Revolutionary days in
Russia and during the post-Revolutionary days, to biologists, these
philosophical-humanistic implications of evolutionism were in the
center of attention.

I think it is not an exaggeration to say that probably this interest is
what made me, if not a biologist, at least an evolutionist. (Dobzhansky
1962–1963, pp. 350–352)

Conversely, religion echoes the progressionism which is confirmed
by the science.

Christianity is basically evolutionistic. It affirms that the meaning of
history lies in the progression from Creation, through Redemption, to
the City of God. (Dobzhansky 1967, p. 112)

What I argue, therefore, is that Dobzhansky was a biological progressionist for a number of reasons, but the chief one of these was

his Christianity. For him, progress and Christianity went hand in hand, and by his own admission he became an evolutionary biologist to find supporting evidence for this worldview. Obviously, all of this predates his reading of Teilhard de Chardin, but when Dobzhansky came upon the Jesuit's writings he found therein just the synthesis to which he had long subscribed. The paradox of Dobzhansky's Teilhardism is a paradox no longer. What would have been surprising would have been an indifference to the *Phenomenon of Man* (Teilhard de Chardin 1955).

THE PROBLEM OF EVIL

We are not quite home yet. There are two large loose ends. On the one hand, we still do not know why Dobzhansky made so much of adaptability as a criterion of progress. It is true that he considered it to be biologically important. But why not a more conventional standard, like complexity or intelligence? On the other hand, we have not yet truly explained Dobzhansky's linking of Christianity and progress. Many, possibly the majority of, Christians, repudiate such a link (Bury 1920). They think progress to be a secular notion, essentially incompatible with the central beliefs of Christianity. The latter speaks of the Fall and of human salvation through Jesus's sacrifice. The tradition—at least, an overwhelming tradition—is that there is nothing we can do without God's help. On our own, whatever our efforts and achievements, we are damned. Christ's grace alone brings triumph. Acceptance of progress, therefore, is a heretical refusal to acknowledge our personally worthless state. Let us not forget that, for all of Dobzhansky's enthusiasm, Teilhard died forbidden to publish his books, out of theological favor with his church.

These two unanswered points are related by a shared solution. Although orthodox theologies stress the significance of Providence, God's working His way in the world, over progress, our working our way in the world, for Dobzhansky there was an overriding theological problem that could be solved only by the invocation of progress, a progress that in some crucial way rested on adaptability. For him, always, especially after the terrors of the Russian Revolution—compounded by the death of his father from syphilis and the fatal choking of his mother on a dry piece of bread—*the* major problem in life was that of reconciling a good God with the all-too-apparent fact of evil. Like his distant relative, Dostoevsky, Dobzhansky was obsessed with finding an answer. And like Dostoevsky—no doubt following Dostoevsky—Dobzhansky found the answer in the fact of freedom. Evil exists, but it is in part a function of human freedom and in part a challenge to freedom, to fight and to overwhelm it. This is the task, and it is in this sense that Christianity incorporates hope of progress, as we free beings try to conquer evil.

What Dobzhansky did was take his religious concerns and read them right back into his science.

> The intellectual stimulation derived from the works of Darwin and other evolutionists was pitted against that arising from reading Dostoevsky, to a lesser extent Tolstoy, and philosophers such as Soloviev and Bergson. Some sort of reconciliation or harmonization seemed necessary. The urgency of finding a meaning of life grew in the bloody tumult of the Russian Revolution, when life became most insecure and its sense least intelligible. (Dobzhansky 1967, p. 1)

And the reconciliation came in a progressivist reading of nature, incorporating a vision where organic evolution is itself progressive, culminating in the organism with the greatest adaptability or freedom, humankind.

> If evolution follows a path which is predestined (orthogenesis), or if it is propelled and guided toward some goal by divine interventions (finalism), then its meaning becomes a tantalizing, and even distressing puzzle. If the universe was designed to advance toward some state of absolute beauty and goodness, the design was incredibly faulty. Why, indeed, should many billions of years be needed to achieve the consummation? The universe could have been created in the state of perfection. Why so many false starts, extinctions, disasters, misery, anguish, and finally the greatest of evils—death? The God of love and mercy could not have planned all this. Any doctrine which regards evolution as predetermined or guided collides head-on with the ineluctable fact of the existence of evil.
>
> Philosophers have struggled with the problem of evil for more than two millennia. Teilhard certainly knew all this, and knew that the only hope for a solution lies in the replacement of predestination by freedom as the mainspring of creation. On the human level, freedom necessarily entails the ability to do evil as well as good. If we can only do the good, or act in only one way, we are not free. We are slaves of necessity. The evolution of the universe must be conceived as having been in some sense a struggle for a gradual emergence of freedom. (Dobzhansky 1967, p. 120)

It is, therefore, in this way that Dobzhansky links progress and adaptability (equated with freedom) as the basis of his case against the immense theological—and personal—problem of evil.

CONCLUSION

Let me put my argument together. I argue that the crucial question for Dobzhansky, from religion as reinforced by his own life experiences, was the problem of evil. This he solved by the undoubted existence of personal freedom, which in some way he equated with adaptability—the ability to respond to life's challenges. (Interestingly, there is a similar equation by Sewall Wright.) All of this leads to human progress which gives meaning to life. Human progress, individually and as a group (a distinction blurred by Dobzhansky), is part of an overall metaphysical progress, ultimately the backbone of Christianity. One's faith in all of this is supported by (what Dobzhansky took to be) the empirical evidence of progress in the biological world. This latter, as with other notions of progress (of which it is indeed a part), centers on adaptability, which at the group level rests on the existence of that genetic variation held always within natural populations.

And as a codicil to this line of arguing, we find that Dobzhansky believes in and worries about the biological counterpart to human evil. These are the individually deleterious genes that populations contain that make possible adaptability. Thus, as we end this discussion, we can nicely speak to another aspect of Dobzhansky's thought that many find paradoxical, namely the fact that—for all his disclaimers—even in his scientific work Dobzhansky keeps harping on the inherently evil side to the natural world:

A species perfectly adapted to its environment may be destroyed by a change in the latter if no hereditary variability is available in the hour of need. Evolutionary plasticity can be purchased only at the ruthlessly dear price of continuously sacrificing some individuals to death from unfavorable mutations. Bemoaning this imperfection of nature has, however, no place in a scientific treatment of this subject. (Dobzhansky 1937, pp. 126–127)

Understanding the significance of this "imperfection of nature" has, however, a major place in a historical treatment of this century's greatest contributor to our understanding of the subject.

REFERENCES

Adams, M. 1980a. "Sergei Chetverikov, the Kol'tsov Institute, and the Evolutionary Synthesis." In E. Mayr and W. B. Provine, eds., *The Evolutionary Synthesis: Perspectives on the Unification of Biology* (Cambridge: Harvard University Press), pp. 242–78.

———. 1980b. "Severtsov and Schmalhausen: Russian Morphology and the Evolutionary Synthesis." In E. Mayr and W. B. Provine, eds., *The Evolutionary Synthesis: Perspectives on the Unification of Biology* (Cambridge: Harvard University Press), pp. 193–225.

———, ed. 1990. *The Wellborn Science: Eugenics in Germany, France, Brazil, and Russia.* New York: Oxford University Press.

Almond, G., M. Chodorow, and R. M. Pearce, eds. 1982. *Progress and Its Discontents.* Berkeley: University of California Press.

Ayala, F. J. 1974. "The Concept of Biological Progress." In Ayala and Dobzhansky 1974, pp. 339–54.

Ayala, F. J., and Th. Dobzhansky, eds. 1974. *Studies in the Philosophy of Biology.* London: Macmillan.

Beatty, J. 1987a. "Dobzhansky and Drift: Facts, Values, and Chance in Evolutionary Biology." In L. Kruger et al., eds., *The Probabilistic Revolution.* Vol. 2, *Ideas in the Sciences.* Cambridge: MIT Press.

———. 1987b. "Weighing the Risks: Stalemate in the Classical/Balance Controversy." *Journal of the History of Biology* 20: 289–319.

Bury, J. B. 1920. *The Idea of Progress: An Inquiry into its Origin and Growth.* Reprinted, New York: Dover, 1955.

Dobzhansky, Th. 1937. *Genetics and the Origin of Species.* New York: Columbia University Press.

———. 1956. *The Biological Basis of Human Freedom.* New York: Columbia University Press.

———. 1962. *Mankind Evolving: The Evolution of the Human Species.* New Haven: Yale University Press.

———. 1962–1963. "The Reminisences of Theodosius Dobzhansky." Typed transcript. 2 parts. Oral History Research Office, Columbia University, New York.

———. 1967. *The Biology of Ultimate Concern.* New York: New American Library.

Fisher, R. A. 1930. *The Genetical Theory of Natural Selection.* Oxford: Oxford University Press.

Haldane, J. B. S. 1924. *Daedalus, or Science and the Future.* London: Kegan Paul, Trench, Trubner.

Hull, D. 1976. "Are Species Really Individuals?" *Systematic Zoology* 25: 174–91.

Lewontin, R. C., J. A. Moore, W. B. Provine, and B. Wallace, eds. 1981.

Dobzhansky's Genetics of Natural Populations I–XLIII. New York: Columbia University Press.

Medawar, P. 1969. *The Art of the Soluble.* Harmondsworth, Mddx.: Penguin.

Nitecki, M., ed. 1988. *Evolutionary Progress.* Chicago: University of Chicago Press.

Provine, W. B. 1986. *Sewall Wright and Evolutionary Biology.* Chicago: University of Chicago Press.

Schmalhausen, I. I. 1949. *Factors of Evolution: The Theory of Stabilizing Selection.* Philadelphia: Blakiston.

Severtsov, A. N. 1929. "Directions of Evolution." *Acta Zoologica,* 10: 59–140.

Teilhard de Chardin, P. 1955. *The Phenomenon of Man.* New York: Harper.

CONTRIBUTORS

MARK B. ADAMS
Department of History and Sociology of Science, SAS
University of Pennsylvania
3440 Market St., Suite 500
Philadelphia, PA 19104-3325

DANIEL A. ALEXANDROV
Sector on Evolutionary History and Theory
St. Petersburg Branch, Institute of the History of Science and
Technology
Russian Academy of Sciences
University Embankment 5
St. Petersburg, 199034 Russia

GARLAND E. ALLEN
Department of Biology
Washington University
St. Louis, MO 63130

JOHN BEATTY
Department of Ecology, Evolution, and Behavior
University of Minnesota
Minneapolis, MN 55455

RICHARD M. BURIAN
Center for the Study of Science in Society
Virginia Polytechnic Institute and State University
Blacksburg, VA 24061-0247

SOPHIA DOBZHANSKY COE
376 St. Ronan St.
New Haven, CN 06511

SCOTT F. GILBERT
Department of Biology
Swarthmore College
Swarthmore, PA 19081

ROBERT E. KOHLER, JR.
Department of History and Sociology of Science, SAS
University of Pennsylvania
3440 Market St., Suite 500
Philadelphia, PA 19104-3325

MIKHAIL KONASHEV
Sector on Evolutionary History and Theory
St. Petersburg Branch, Institute of the History of Science and
Technology
Russian Academy of Sciences
University Embankment 5
St. Petersburg, 199034 Russia

NIKOLAI L. KREMENTSOV
Sector on Evolutionary History and Theory
St. Petersburg Branch, Institute of the History of Science and
Technology
Russian Academy of Sciences
University Embankment 5
St. Petersburg, 199034 Russia

COSTAS B. KRIMBAS
Department of Genetics
Agricultural University of Athens
Iera Odos 75
Athens 118 55, Greece

DIANE B. PAUL
Department of Political Science
University of Massachusetts, Boston, Harbor Campus
Boston, MA 02125

WILLIAM B. PROVINE
Section of Ecology and Systematics
Corson Hall, Cornell University
Ithaca, NY 14853

MICHAEL RUSE
Department of Philosophy
University of Guelph
Guelph, Ontario, Canada N1G 2W1

CHARLES E. TAYLOR
Department of Biology
405 Hilgard Ave.
University of California, Los Angeles (UCLA)
Los Angeles, CA 90024-1606

BRUCE WALLACE
Department of Biology
Virginia Polytechnic Institute and State University
Blacksburg, VA 24061-0406

Milton Keynes UK
Ingram Content Group UK Ltd.
UKHW021817250823
427506UK00006B/167